Uterine Endometrial Function

Hideharu Kanzaki
Editor

Uterine Endometrial Function

Springer

Editor
Hideharu Kanzaki
Department of Obstetrics and Gynecology
Kansai Medical University
Osaka, Japan

ISBN 978-4-431-55970-2 ISBN 978-4-431-55972-6 (eBook)
DOI 10.1007/978-4-431-55972-6

Library of Congress Control Number: 2016938798

Printed on acid-free paper

This Springer imprint is published by Springer Nature
The registered company is Springer Japan KK

Preface

This book focuses, from multiple perspectives, on uterine endometrial function and its receptivity for the fertilized ovum. Morphological and functional uterine endometrial changes are primarily controlled by ovarian steroidal hormones and are regulated by many secondary messenger molecules. Despite nearly 40 years of experience with in vitro fertilization (IVF), the rate of successful implantations remains low. Steady progress has been made in understanding of the fertilization process, and the development of the intracytoplasmic sperm injection technique has remarkably improved the fertilization rate in vitro. As well, advances in culture medium and equipment enable the fertilized ovum to develop up to the pre-implantation blastocyst stage. On the other hand, because of the limitations of experimental models, our understanding of the implantation process is still greatly limited, but recent molecular and genetic studies have gradually been clarifying the details. Endometrial receptivity results from an orchestrated interplay between the fertilized ovum and the maternal uterine endometrium; and its receptive status, known as the window of implantation, is reached only briefly in the mid-luteal phase as a result of a harmonized reciprocal relationship.

Abnormal endometrial receptivity is, therefore, one of the factors contributing to reduced reproductive potential in women and is the greatest challenge in infertility treatment, remaining the last intractable problem in IVF practice. This book provides a comprehensive overview of the latest advances in endometrial function and paves the way for innovative treatments and drug development for infertility. The chapters cover a variety of topics including the mechanism of menstruation, animal models, parameters for assessing endometrial receptivity, mechanism of angiogenesis, the role of immune cells, epigenetic regulation and stem/progenitor cells, and related information. The book will provide an important reference for researchers on the biology of reproduction as well as for clinicians and technicians in the field of reproductive medicine.

Osaka, Japan Hideharu Kanzaki

Contents

Chapter 1
ERα Signal Pathways Regulating *Bcl-2* Transcription in Human Endometrial Glands

Yoshinori Otsuki

Abstract Although it has been shown that the cyclic transition of endometrium is genetically controlled, we still do not know the specific genes involved. It has been suggested that the differences in estrogen-induced cell proliferation and apoptosis are involved in the different phenotypes between mammary gland and endometrium. In the glandular cells of human endometrium, the expression of Bcl-2 is increased at proliferative phases but not at late secretory through menstrual phases. The disappearance of Bcl-2 expression in glandular cells at late secretory phase is consistent with the appearance of apoptotic cells at the same phase. The proliferative phase-specific expression of Bcl-2 in glandular cells is regulated by the binding of C-Jun to its motifs in the promoter region. Furthermore, the menstrual cyclic expression of Bcl-2 is regulated either by the interaction of ERα with C-Jun that binds to its motifs in the *Bcl-2* gene or by direct binding of ERα to ERE in the *C-Jun* promoter. The discovery of new mechanisms downstream of ERα will help pave the way for further understanding of human endometrial transition and function.

Keywords Endometrium • ERα • AP-1 • C-Jun • Bcl-2

1.1 Introduction

Endometrium and mammary gland are two major estrogen-responsive tissues of most mammals, including humans. It is now generally accepted [1, 2] that the responsiveness of endometrium and mammary gland to estrogen is genetically controlled, with marked quantitative variations in addition to similarities. It has been suggested that such variations in responsiveness to estrogen appear to involve differences in estrogen-induced cell proliferation and apoptosis [3]. Based on the current retrieving numbers of articles, however, research on estrogen receptor alpha (ERα) in the uterus (endometrium and ERα) is far less active than that in mammary

Y. Otsuki (✉)
Department of Anatomy and Biology, Osaka Medical College, 2-7 Daigaku-machi, Takatsuki, Osaka 569-8686, Japan
e-mail: an1001@art.osaka-med.ac.jp

H. Kanzaki (ed.), *Uterine Endometrial Function*,
DOI 10.1007/978-4-431-55972-6_1

gland (breast and ERα); and similar tendency could also be found in research about the ERα-related Bcl-2. When addressing endometrium-specific responsiveness to estrogen in human, therefore, it is essential to tentatively include the ERα-related research conducted on normal or breast cancer cells, including experimental animals.

1.2 Antiapoptotic Bcl-2

Bcl-2, a 25-kD protein, is the first identified member of a family of apoptotic regulators [4, 5]. Bcl-2 family is comprised of both proapoptotic (e.g., Bax, Bak, and Bid) and antiapoptotic (e.g., Bcl-2, Bcl-x_L, and Mcl-1) members. The members of the Bcl-2 family share one or more of the four characteristic domains of homology entitled the Bcl-2 homology (BH) domains, which are known to be crucial for function. Bcl-2 family acts as a checkpoint in mitochondria. Interactions between and relative ratios of proapoptotic and antiapoptotic Bcl-2 family members are key determinants of cell survival [5].

Bcl-2 is found in several subcellular locations including the outer mitochondrial, outer nuclear, and endoplasmic reticulum membranes [6–8] of endometrial glandular cells [6]. It is generally accepted that mitochondrial dysfunction leads to decrease in the mitochondrial membrane potential [9] and the release of cytochrome *c* from the intermembrane space [10]. Most importantly, unlike oncogenes studied previously, Bcl-2 functions in preventing programmed cell death other than promoting proliferation [11]. For example, while the Bcl-2-deficient T cells exhibit accelerated cell cycle progression [12], the Bcl-2 overexpressing peripheral T cells show delayed entry to S phase [13, 14].

1.3 Cyclic Expression of Bcl-2 in Human Endometrial Glands

Extensive research demonstrates that expression of Bcl-2 is regulated in both a tissue-and a time-specific manner [4, 15]. We and other researchers have shown that there is cyclic expression of Bcl-2 in the glandular cells of the human endometrium, in which levels of Bcl-2 increase gradually during the proliferative phase and disappear soon after the beginning of secretory phase in the menstrual cycle [15]. The stroma, surface lining epithelium and arterial vessels also display cyclic variations in Bcl-2 expression. This cyclic expression of Bcl-2 gene is similar to that of the ERα, and the disappearance of Bcl-2 expression in glandular cells at late secretory phase is consistent with the appearance of apoptotic cells in the same phase [16]. These results are strongly related to hormone-dependent regulation of Bcl-2 expression [17].

1.4 Transcription of Bcl-2 in ERα-Responsive Cancer Cells

More and more information has become available concerning the transcriptional control of *Bcl-2* gene [18, 19]. Perillo et al. [20] have shown that hormone prevention of apoptosis in breast cancer MCF-7 cells is strictly related to Bcl-2 upregulation. The Bcl-2 expression is induced via two estrogen-responsive elements located within its coding region, rather than the promoter. Other researchers [21] have demonstrated that the increased reporter activity is detectable after co-transfection of a vector expressing Sp1 and a reporter plasmid containing the Sp1-binding site from the *Bcl-2* 5′ promoter region. Interestingly, it is also shown that ATF-1, a member of the activating protein-1 (AP-1) family, is also involved in regulation of Bcl-2 expression.

Concerning the downregulation of Bcl-2, it is reported that DNA damage-binding protein complex (DDB), composed of two subunits, DDB1 and DDB2, functions as a transcriptional repressor for Bcl-2 [22]. They suggest that DDB1 and DDB2 cooperate to repress *Bcl-2* transcription, independent of p53 pathways by inhibiting *Bcl-2* transcription and promoting Bcl-2 degradation via the ubiquitin-proteasome pathway. In addition, DDB2 recognizes and binds to the *Bcl-2* P1 promoter, and histone deacetylase 1 (HDAC1) is recruited through the DDB1 subunit associated with DDB2 to deacetylate histone H3K9 across *Bcl-2* regulatory regions, which results in suppressed *Bcl-2* transcription [22], indicating critical roles of epigenetic regulation [23].

Emerging evidence indicates that epigenetic regulations are crucially important to many aspects of gene transcription. It is reported [24] that in *Bcl-2* promoter, a DNA secondary structure formed in G-rich region called G-quadruplex [25, 26] is intimately related with its transcription activity. On the other hand in the involuting mammary gland, Llobet-Navas et al. [27] have shown that microRNA cluster miR-424(322)/503 functions to downregulate Bcl-2. Another microRNA miR-21 is also involved in downregulation of Bcl-2 in MCF-7 cells [28]. Clearly, microRNAs play a crucial role on expression of various genes in endometrial cancer cells [29].

1.5 Regulation of Transcription by ERα

By regulating various target genes, the ERα signaling pathways play critical physiologic roles, including not only the control of reproduction and development but also the functions of the central nervous, skeletal, and cardiovascular systems [30]. On the other hand, the estradiol and ER play crucial roles in the development and growth of a variety of cancers, most notably breast cancer [31] and uterine carcinoma [32]. Although the molecular mechanisms of ER and estrogen action are relatively well understood, only some target genes, such as those encoding the progesterone receptor, pS2, vitellogenin, cathepsin D1, and estrogen-responsive finger protein [30, 33], have been identified containing the consensus estrogen-responsive element (ERE). Each ERE is composed of two hexanucleotide half-sites separated by three nucleotides (GGTCAnnnTGACC). The sequences of the half-sites and the number are key determinants of the specificity of ER interaction [34]. In addition to the homo- or heterodimers of ERα and ERβ [35], a greater complexity was attributed to the coregulators in an ER/DNA-binding complex [36, 37].

Other E2-responsive genes are regulated by ER via protein-protein interaction but without apparent ERE [33, 38]. The effects are either mediated through coregulators associated with a multisubunit DNA-binding complex, including the TATA-binding protein (TBP) and RNA polymerase II (Pol II), or exerted by modulation of other transcription factors that bind to their responsive elements. One typical example of the latter case is the ER-stimulated transcription from the promoter containing a motif called activator protein-1 (AP-1), the cognate binding site for AP-1 transcription factors.

It should be noted that the function of ERα is closely related to its state of phosphorylation [39] and its cross talk with various kinase signaling pathways [40]. There is ample evidence showing that regulation of ERα-mediated transcription by cross talk with PI3K-AKT and Raf-MEK-ERK pathways has critical impact on breast cancer cell survival [40, 41], and Bcl-2 is one of the important genes affected [41]. It is plausible to accept that various kinase signaling pathways play indespensable physiological roles in human endometrium [42].

1.6 Cyclic Expression of C-Jun in Human Endometrial Glands

The expression pattern of C-Jun in endometrial glands is similar to that of the Bcl-2 described previously. Intense immunoreactivity of C-Jun is detected in the endometrial glandular cells during the proliferative phase, whereas its immunoreactivity is markedly decreased in the secretory glandular cells [17, 43]. The expression pattern of C-Jun is consistent with that reported by Salmi et al. [44, 45]. It is exhibited that there is an intense staining and increased amount of expression during the proliferative phase and a decreased amount of expression and immunoreactivity in nuclei of the glandular cells before menstruation. The importance of C-Jun is further confirmed during exploration of the mechanism of estrogen-induced growth of normal endometrium. Shiozawa et al. [46] have reported that upregulation of C-Jun and other proteins is inducible following estradiol (E2) treatment of cultured normal endometrial glandular cells.

1.7 Regulation of Transcription by C-Jun

The activity of C-Jun, one of the major components of AP-1, is regulated at the transcriptional level, as well as posttranslationally. Changes in the phosphorylation state of C-Jun are required to generate transactivation potential [47, 48]. Thus, the same stimuli that induce C-Jun expression also trigger its phosphorylation at Ser63 and Ser73 in the N-terminal domain [49], which is required for it to become transcriptionally active. The phosphorylation of these residues is considered to be mediated by the isoforms of C-Jun N-terminal kinase (JNK). It has also been reported that C-Jun becomes phosphorylated at other residues proximal to the DNA-binding domain. Thr239, Ser243, and Ser249 were reported to be phosphorylated by GSK [50], and Thr231 and Ser249 by CK2 in vitro [51], and to inhibit the binding of C-Jun to DNA. All the four residues are situated within the same tryptic peptide whose phosphorylation is inhibited when either HeLa or human osteosarcoma MG63 cells are stimulated with the tumor-promoting phorbol ester TPA. Thus, it has generally accepted that the activation of C-Jun requires the phosphorylation of Ser63/Ser73, as well as the dephosphorylation of one or several C-terminal sites.

1.8 Transcription of Bcl-2 Is Controlled by ERα Through C-Jun

It was generally accepted [33] that the transcription of C-Jun is controlled by estrogen through the binding of ERα to ERE, based on the fact that the expression of C-Jun in the presence of estrogen could be increased in either the rodent uterine or breast and endometrial malignant cells. However, a detailed survey of the reports showed that the existence of ERE has never been confirmed in human C-Jun, rather a suggestion based on studies of the rodent homolog [52]. Our results [17] showed that although all the three incomplete ERE in human *C-Jun* promoter could stably bind to the recombinant human ERα, only ERE-A could bind to the ERα in nuclear proteins prepared from glandular cells at the proliferative rather than secretory phase. The proliferative phase-specific binding is consistent with the cyclic expression of C-Jun in our previous report [43], which has been confirmed by many other researchers [45, 53].

In addition to the direct action of ERα, coactivators, such as SRC-1, CBP/p300, and pCAF, or corepressors such as N-CoR and SMRT could be found in an ER-DNA-binding complex [36, 37]. It has been reported that while the expression levels of coactivators such as SRC-1 and CBP/p300 [54] are elevated during the proliferative phase, the decrease in the amount of corepressors N-Cor and SMRT [55] was observed during menstruation. Intriguingly, it has been reported that the C-Jun transcription can be directly stimulated by its own gene product [56], which may confer its early, intensive, and prolonged responsiveness.

We have reported that C-Jun in glandular cells of the proliferative endometrium could bind to its motifs in the human *Bcl-2* promoter [43]. In addition we showed that the same motifs could also bind to ERα from glandular nuclear proteins in a proliferative phase-specific manner. Therefore, it is suggested that ERα could control the menstrual cyclic transcription of *Bcl-2* via either direct binding to the incomplete ERE in the *C-Jun* promoter or through protein-protein interaction with C-Jun on AP-1 sites in the *Bcl-2* promoter. The mechanism is novel because many other factors in this protein-DNA or protein-protein complex could be involved in a subtle regulation of both C-Jun and Bcl-2 in the normal human endometrium.

1.9 Conclusion

Marked variation in ERα-mediated responses has been documented for many tissues including endometrium, mammary gland, and other tissues based on the differences in genetic background [3]. Although it is now accepted that genetics explain a major part of the variations in ERα-mediated responses, we still do not know the specific genes involved in different phenotypes between endometrium and mammary gland. Elucidating the regulation of specific target genes such as *C-Jun* and *Bcl-2* in the estrogen-ERα signaling pathways would be important for further understanding of the cyclic transition and tumorigenesis in human endometrium.

Acknowledgments The present study is supported in part by a Grant-in-Aid for General Scientific Research from the Ministry of Education, Culture, Sports, and Technology of Japan (no. 10671576). The author is very grateful to Dr. Zhonglian Li for his contribution.

References

1. Wall EH, Case LK, Hewitt SC, Nguyen-Vu T, Candelaria NR, Teuscher C, et al. Genetic control of ductal morphology, estrogen-induced ductal growth, and gene expression in female mouse mammary gland. Endocrinology. 2014;155(8):3025–35.
2. Roper RJ, Griffith JS, Lyttle CR, Doerge RW, McNabb AW, Broadbent RE, et al. Interacting quantitative trait loci control phenotypic variation in murine estradiol-regulated responses. Endocrinology. 1999;140(2):556–61.
3. Wall EH, Hewitt SC, Case LK, Lin CY, Korach KS, Teuscher C. The role of genetics in estrogen responses: a critical piece of an intricate puzzle. FASEB J. 2014;28(12):5042–54.
4. Chao DT, Korsmeyer SJ. BCL-2 family: regulators of cell death. Annu Rev Immunol. 1998;16:395–419.
5. Korsmeyer SJ. BCL-2 gene family and the regulation of programmed cell death. Cancer Res. 1999;59(7 Suppl):1693s–700.
6. Akao Y, Otsuki Y, Kataoka S, Ito Y, Tsujimoto Y. Multiple subcellular localization of bcl-2: detection in nuclear outer membrane, endoplasmic reticulum membrane, and mitochondrial membranes. Cancer Res. 1994;54(9):2468–71.

7. Krajewski S, Tanaka S, Takayama S, Schibler MJ, Fenton W, Reed JC. Investigation of the subcellular distribution of the bcl-2 oncoprotein: residence in the nuclear envelope, endoplasmic reticulum, and outer mitochondrial membranes. Cancer Res. 1993;53(19):4701–14.
8. Tsujimoto Y, Cossman J, Jaffe E, Croce CM. Involvement of the bcl-2 gene in human follicular lymphoma. Science. 1985;228(4706):1440–3.
9. Schlesinger PH, Gross A, Yin XM, Yamamoto K, Saito M, Waksman G, et al. Comparison of the ion channel characteristics of proapoptotic BAX and antiapoptotic BCL-2. Proc Natl Acad Sci U S A. 1997;94(21):11357–62.
10. Kluck RM, Bossy-Wetzel E, Green DR, Newmeyer DD. The release of cytochrome c from mitochondria: a primary site for Bcl-2 regulation of apoptosis. Science. 1997;275 (5303):1132–6.
11. Korsmeyer SJ. Bcl-2 initiates a new category of oncogenes: regulators of cell death. Blood. 1992;80(4):879–86.
12. Linette GP, Li Y, Roth K, Korsmeyer SJ. Cross talk between cell death and cell cycle progression: BCL-2 regulates NFAT-mediated activation. Proc Natl Acad Sci U S A. 1996;93(18):9545–52.
13. O'Reilly LA, Huang DC, Strasser A. The cell death inhibitor Bcl-2 and its homologues influence control of cell cycle entry. EMBO J. 1996;15(24):6979–90.
14. Vairo G, Innes KM, Adams JM. Bcl-2 has a cell cycle inhibitory function separable from its enhancement of cell survival. Oncogene. 1996;13(7):1511–9.
15. Otsuki Y, Misaki O, Sugimoto O, Ito Y, Tsujimoto Y, Akao Y. Cyclic bcl-2 gene expression in human uterine endometrium during menstrual cycle. Lancet. 1994;344(8914):28–9.
16. Dualan R, Brody T, Keeney S, Nichols AF, Admon A, Linn S. Chromosomal localization and cDNA cloning of the genes (DDB1 and DDB2) for the p127 and p48 subunits of a human damage-specific DNA binding protein. Genomics. 1995;29(1):62–9.
17. Li ZL, Ueki K, Kumagai K, Araki R, Otsuki Y. Regulation of bcl-2 transcription by estrogen receptor-alpha and c-Jun in human endometrium. Med Mol Morphol. 2014;47(1):43–53.
18. Svotelis A, Bianco S, Madore J, Huppe G, Nordell-Markovits A, Mes-Masson AM, et al. H3K27 demethylation by JMJD3 at a poised enhancer of anti-apoptotic gene BCL2 determines ERalpha ligand dependency. EMBO J. 2011;30(19):3947–61.
19. Smith MD, Ensor EA, Coffin RS, Boxer LM, Latchman DS. Bcl-2 transcription from the proximal P2 promoter is activated in neuronal cells by the Brn-3a POU family transcription factor. J Biol Chem. 1998;273(27):16715–22.
20. Perillo B, Sasso A, Abbondanza C, Palumbo G. 17beta-estradiol inhibits apoptosis in MCF-7 cells, inducing bcl-2 expression via two estrogen-responsive elements present in the coding sequence. Mol Cell Biol. 2000;20(8):2890–901.
21. Dong L, Wang W, Wang F, Stoner M, Reed JC, Harigai M, et al. Mechanisms of transcriptional activation of bcl-2 gene expression by 17beta-estradiol in breast cancer cells. J Biol Chem. 1999;274(45):32099–107.
22. Zhao R, Han C, Eisenhauer E, Kroger J, Zhao W, Yu J, et al. DNA damage-binding complex recruits HDAC1 to repress Bcl-2 transcription in human ovarian cancer cells. Mol Cancer Res. 2014;12(3):370–80.
23. Banno K, Yanokura M, Iida M, Masuda K, Aoki D. Carcinogenic mechanisms of endometrial cancer: involvement of genetics and epigenetics. J Obstet Gynaecol Res. 2014;40(8):1957–67.
24. Sun H, Xiang J, Shi Y, Yang Q, Guan A, Li Q, et al. A newly identified G-quadruplex as a potential target regulating Bcl-2 expression. Biochim Biophys Acta. 2014;1840(10):3052–7.
25. Onyshchenko MI, Gaynutdinov TI, Englund EA, Appella DH, Neumann RD, Panyutin IG. Quadruplex formation is necessary for stable PNA invasion into duplex DNA of BCL2 promoter region. Nucleic Acids Res. 2011;39(16):7114–23.
26. Lin J, Hou JQ, Xiang HD, Yan YY, Gu YC, Tan JH, et al. Stabilization of G-quadruplex DNA by C-5-methyl-cytosine in bcl-2 promoter: implications for epigenetic regulation. Biochem Biophys Res Commun. 2013;433(4):368–73.

27. Llobet-Navas D, Rodriguez-Barrueco R, Castro V, Ugalde AP, Sumazin P, Jacob-Sendler D, et al. The miR-424(322)/503 cluster orchestrates remodeling of the epithelium in the involuting mammary gland. Genes Dev. 2014;28(7):765–82.
28. Wickramasinghe NS, Manavalan TT, Dougherty SM, Riggs KA, Li Y, Klinge CM. Estradiol downregulates miR-21 expression and increases miR-21 target gene expression in MCF-7 breast cancer cells. Nucleic Acids Res. 2009;37(8):2584–95.
29. Klinge CM. miRNAs regulated by estrogens, tamoxifen, and endocrine disruptors and their downstream gene targets. Mol Cell Endocrinol. 2015. doi:10.1016/j.mce.2015.01.035.
30. Muramatsu M, Inoue S. Estrogen receptors: how do they control reproductive and nonreproductive functions? Biochem Biophys Res Commun. 2000;270(1):1–10.
31. Lawson JS, Field AS, Champion S, Tran D, Ishikura H, Trichopoulos D. Low oestrogen receptor alpha expression in normal breast tissue underlies low breast cancer incidence in Japan. Lancet. 1999;354(9192):1787–8.
32. Li JJ, Li SA. Causation and prevention of solely estrogen-induced oncogenesis: similarities to human ductal breast cancer. Adv Exp Med Biol. 2003;532:195–207.
33. Rollerova E, Urbancikova M. Intracellular estrogen receptors, their characterization and function (Review). Endocr Regul. 2000;34(4):203–18.
34. Naar AM, Boutin JM, Lipkin SM, Yu VC, Holloway JM, Glass CK, et al. The orientation and spacing of core DNA-binding motifs dictate selective transcriptional responses to three nuclear receptors. Cell. 1991;65(7):1267–79.
35. Johnston SD, Liu X, Zuo F, Eisenbraun TL, Wiley SR, Kraus RJ, et al. Estrogen-related receptor alpha 1 functionally binds as a monomer to extended half-site sequences including ones contained within estrogen-response elements. Mol Endocrinol. 1997;11(3):342–52.
36. Robyr D, Wolffe AP, Wahli W. Nuclear hormone receptor coregulators in action: diversity for shared tasks. Mol Endocrinol. 2000;14(3):329–47.
37. Jenster G. Coactivators and corepressors as mediators of nuclear receptor function: an update. Mol Cell Endocrinol. 1998;143(1-2):1–7.
38. Katzenellenbogen JA, Katzenellenbogen BS. Nuclear hormone receptors: ligand-activated regulators of transcription and diverse cell responses. Chem Biol. 1996;3(7):529–36.
39. Anbalagan M, Rowan BG. Estrogen receptor alpha phosphorylation and its functional impact in human breast cancer. Mol Cell Endocrinol. 2015. doi:10.1016/j.mce.2015.01.016.
40. Saini KS, Loi S, de Azambuja E, Metzger-Filho O, Saini ML, Ignatiadis M, et al. Targeting the PI3K/AKT/mTOR and Raf/MEK/ERK pathways in the treatment of breast cancer. Cancer Treat Rev. 2013;39(8):935–46.
41. Bratton MR, Duong BN, Elliott S, Weldon CB, Beckman BS, McLachlan JA, et al. Regulation of ERalpha-mediated transcription of Bcl-2 by PI3K-AKT crosstalk: implications for breast cancer cell survival. Int J Oncol. 2010;37(3):541–50.
42. Fabi F, Asselin E. Expression, activation, and role of AKT isoforms in the uterus. Reproduction. 2014;148(5):R85–95.
43. Li ZL, Abe H, Ueki K, Kumagai K, Araki R, Otsuki Y. Identification of c-Jun as bcl-2 transcription factor in human uterine endometrium. J Histochem Cytochem. 2003;51 (12):1601–9.
44. Salmi A, Carpen O, Rutanen E. The association between c-fos and c-jun expression and estrogen and progesterone receptors is lost in human endometrial cancer. Tumour Biol. 1999;20(4):202–11.
45. Salmi A, Rutanen FM. C-fos and c-jun expression in human endometrium and myometrium. Mol Cell Endocrinol. 1996;117(2):233–40.
46. Shiozawa T, Miyamoto T, Kashima H, Nakayama K, Nikaido T, Konishi I. Estrogen-induced proliferation of normal endometrial glandular cells is initiated by transcriptional activation of cyclin D1 via binding of c-Jun to an AP-1 sequence. Oncogene. 2004;23(53):8603–10.
47. Chang L, Karin M. Mammalian MAP kinase signalling cascades. Nature. 2001;410 (6824):37–40.

48. Leppa S, Saffrich R, Ansorge W, Bohmann D. Differential regulation of c-Jun by ERK and JNK during PC12 cell differentiation. EMBO J. 1998;17(15):4404–13.
49. Davis RJ. Signal transduction by the JNK group of MAP kinases. Cell. 2000;103(2):239–52.
50. Boyle WJ, Smeal T, Defize LH, Angel P, Woodgett JR, Karin M, et al. Activation of protein kinase C decreases phosphorylation of c-Jun at sites that negatively regulate its DNA-binding activity. Cell. 1991;64(3):573–84.
51. Lin A, Frost J, Deng T, Smeal T, al-Alawi N, Kikkawa U, et al. Casein kinase II is a negative regulator of c-Jun DNA binding and AP-1 activity. Cell. 1992;70(5):777–89.
52. Hyder SM, Nawaz Z, Chiappetta C, Yokoyama K, Stancel GM. The protooncogene c-jun contains an unusual estrogen-inducible enhancer within the coding sequence. J Biol Chem. 1995;270(15):8506–13.
53. Maldonado V, Castilla JA, Martinez L, Herruzo A, Concha A, Fontes J, et al. Expression of transcription factors in endometrium during natural cycles. J Assist Reprod Genet. 2003;20 (11):474–81.
54. Shiozawa T, Shih HC, Miyamoto T, Feng YZ, Uchikawa J, Itoh K, et al. Cyclic changes in the expression of steroid receptor coactivators and corepressors in the normal human endometrium. J Clin Endocrinol Metab. 2003;88(2):871–8.
55. Wieser F, Schneeberger C, Hudelist G, Singer C, Kurz C, Nagele F, et al. Endometrial nuclear receptor co-factors SRC-1 and N-CoR are increased in human endometrium during menstruation. Mol Hum Reprod. 2002;8(7):644–50.
56. Angel P, Hattori K, Smeal T, Karin M. The jun proto-oncogene is positively autoregulated by its product, Jun/AP-1. Cell. 1988;55(5):875–85.

Chapter 2
Uterine Receptivity in Mouse Embryo Implantation

Yasushi Hirota

Abstract A competent blastocyst and a receptive uterus are two critical components for successful embryo implantation. Currently, mouse models are the most powerful tools to understand mechanisms by which acquisition of uterine receptivity takes place. Based on the previous studies performed by us and others, pre-receptive stromal proliferation and epithelial differentiation regulated by ovarian hormones, which we call endometrial proliferation-differentiation switching (PDS), can be a potent marker of uterine receptivity. Molecular interactions between the uterus and the blastocysts, which are followed by the acquisition of uterine receptivity, allow the subsequent implantation processes such as attachment reaction and decidualization. This chapter shows detailed molecular mechanisms for successful implantation, focusing on uterine receptivity and referring to the mouse in vivo evidence.

Keywords Blastocyst implantation • Uterine receptivity • Ovarian hormones • Proliferation-differentiation switching • Cytokines • Growth factors

2.1 Introduction

Pregnancy is constituted by a series of processes such as ovulation, fertilization, implantation, feto-placental growth, and parturition. Each process is strictly coordinated and essential for successful pregnancy. Implantation, a process of the first embryo-maternal encounter, consists of the following three steps: apposition, adhesion, and invasion of the embryo. Successful implantation is the result of appropriate molecular communications between the uterus and the blastocyst during these steps. Animal studies, especially mouse studies, have been often used in implantation research [1, 2]. Especially, recent studies using genetically modified mice provide us valuable information in the field of implantation research,

---

Y. Hirota (✉)
Department of Obstetrics & Gynecology, Graduate School of Medicine, The University of Tokyo, 7-3-1 Hongo, Bunkyo-ku, Tokyo 113-8655, Japan
e-mail: yhirota-tky@umin.ac.jp

H. Kanzaki (ed.), *Uterine Endometrial Function*,
DOI 10.1007/978-4-431-55972-6_2

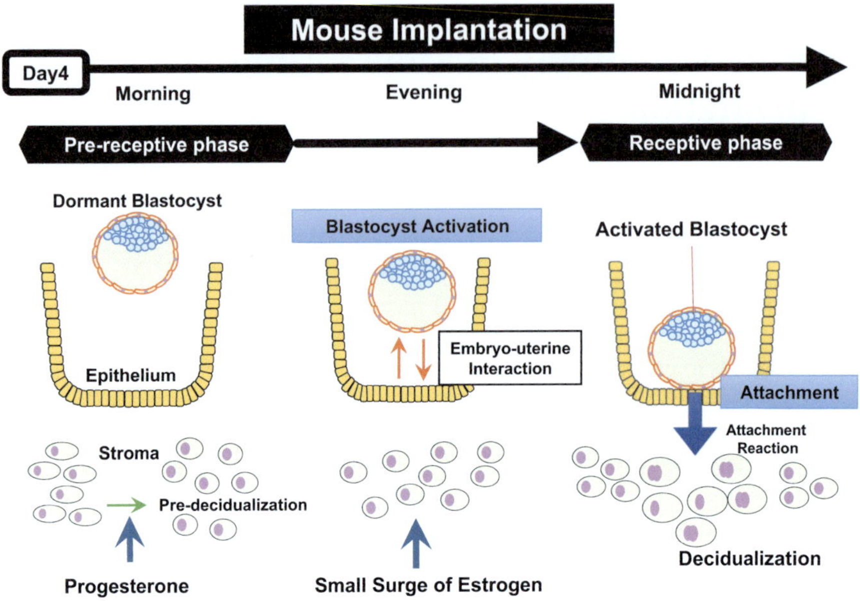

Fig. 2.1 Sequential processes of embryo implantation

and the current concepts in embryo implantation have been primarily proposed by mouse studies.

There are two essential components. One is an implantation-competent blastocyst because poor embryo quality must be one of the major causes of implantation failure [3]. The other is uterine receptivity defined as a capacity to accommodate the competent blastocyst in the uterus [1, 2]. The uterus with this capacity demonstrates a proper endometrial preparation with stromal proliferation and epithelial differentiation stimulated by ovarian steroids in advance before embryo-uterine interactions (Fig. 2.1). In this process, the stroma shows progesterone (P_4)-dependent morphological changes called "pre-decidualization" (Fig. 2.1) [4]. The small spike of ovarian estradiol-17β (E_2) is followed by an acquisition of the endometrial status, and then, the embryo can possess adhesion activity. Thus, the uterus enters into the receptive phase. It is speculated that endometrium-derived factors activate the dormant blastocyst to provide the capacity of implantation, and this concept is called "blastocyst activation" (Fig. 2.1) [1, 2]. Blastocyst adhesion onto the uterus induces an endometrial attachment reaction, in which stromal cells around the blastocyst start to differentiate concurrently with polyploid formation, which is called "decidualization" (Fig. 2.1) [4]. The receptive phase of the uterus is transient, and unless the blastocyst adhesion occurs, the endometrium enters into the refractory phase, when any competent blastocysts can never adhere to the endometrium. Thus, the endometrium enables blastocysts to adhere to itself in the limited period, which is known as "implantation window" (Fig. 2.1) [4]. This sequence of events is fundamental to initiation of implantation.

2.2 Embryo Implantation in Mice and Humans

Timeline and hormonal conditions in the peri-implantation period seem similar between mice and humans (Fig. 2.2), and therefore, mouse studies make us know how embryo implantation occurs [1, 2]. In mice, vaginal plug is observed in the morning on the next day of ovulation and mating, which is defined as day 1 of pregnancy. Luminal epithelium strongly proliferates and the uterus looks swollen under the influence of E_2 surge. On day 3 of pregnancy, newly formed corpora lutea start to secrete P_4. P_4 is completely dominant by day 4 morning when heightened P_4 makes endometrial stromal cells proliferate, called pre-decidualization (Fig. 2.2), and this phenomenon is similarly observed in humans. At the same time, the luminal epithelium declines to proliferate and differentiates for attachment reaction to the blastocyst. Late on day 4 morning, small E_2 spike occurs as a starting signal of implantation (Fig. 2.2). This E_2 surge induces stromal edema and luminal closure placing the blastocyst in close apposition with the luminal epithelium and makes

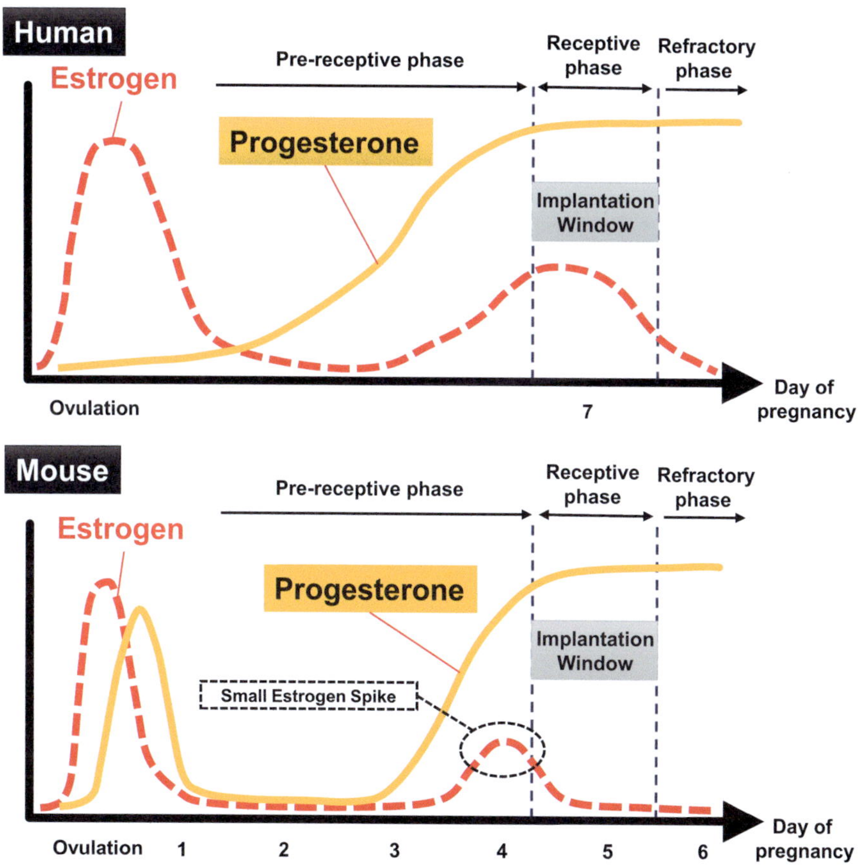

Fig. 2.2 Hormonal condition in early pregnancy of humans and mice

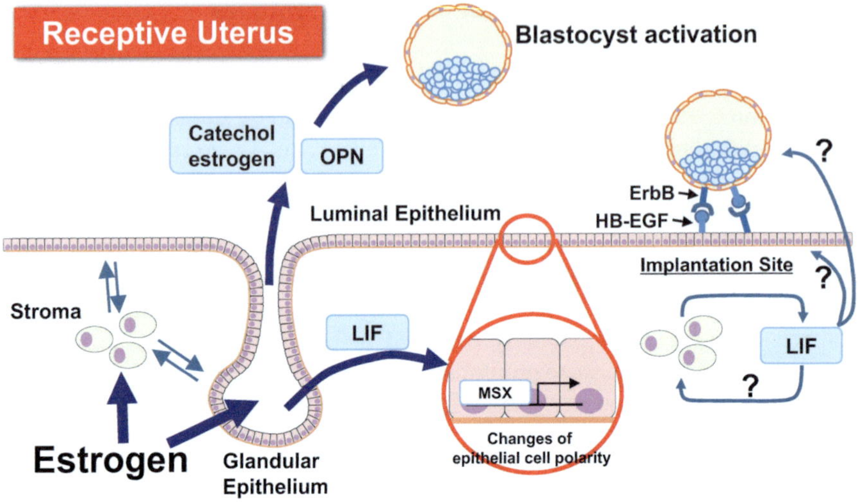

Fig. 2.3 Molecular interactions in the receptive uterus

the uterus produce blastocyst activators such as catechol estrogen and osteopontin (OPN) [5, 6] (Fig. 2.3). It is followed by an intimate adherence of the blastocyst trophectoderm to the luminal epithelium, marking the first apparent sign of implantation on day 4 night (2200–2400 h). Immediately after the implantation, stromal cells surrounding the blastocyst start differentiation, change their stromal morphology into epithelioid type with polyploidy, and form a new layer around the embryo. This process is known as decidualization. The attachment reaction coincides with an increased stromal vascular permeability at the site of the blastocyst. Embryo-derived trophoblast cells invade into the endometrium, and finally, embryo implantation is completed [1, 2].

As described above, the current concepts in the embryo implantation primarily arise from mouse studies. Because observations of blastocyst activation, attachment reaction, and decidualization (not pre-decidualization) are technically and ethically difficult to be performed in humans, mouse models are the most powerful approaches to understand embryo implantation and are worldwide applied in the current research of reproduction.

2.3 Roles of Ovarian Hormones: Estradiol-17β (E_2) and Progesterone (P_4)

E_2 and P_4 play crucial roles throughout pregnancy. The following two processes under the control of ovarian steroids are needed for successful implantation: preparation of endometrial proliferation and differentiation and proper embryo-uterine cross talk. In the pre-receptive phase, the endometrium must possess specific differentiation status in which luminal epithelium eliminates proliferation and subluminal stroma starts to proliferate under the P_4-dominant hormonal condition. Then, a small spike of E_2 occurs just before the receptive phase. This nidatory E_2 with continuous effects of P_4 provides starting signals to the uterus for embryo-uterine communications. Dormant blastocyst is activated by E_2-derived uterine factors, and the uterus turns to be receptive. Thus, the implantation-competent blastocyst as well as the receptive uterus is prepared through the molecular communications between the embryo and uterus under the influence of ovarian hormones [1, 2]. To clarify the molecular and cellular contribution to these processes, the roles of P_4, a "hormone of pregnancy," in implantation should definitely be understood first. In the next section, I describe how P_4 signaling regulates the endometrial differentiation and proliferation in the pre-receptive phase.

2.4 Proliferation-Differentiation Switching (PDS) via P_4 Signaling in the Endometrium

P_4 acts through P_4 receptor (PR), a nuclear receptor, transcriptionally governing P_4-responsive genes and triggering critical pathways for each pregnancy event including ovulation and implantation [2, 4, 7, 8]. The genetic modification of PR in female mice has provided many insights into the role of P_4 during pregnancy. PR-deficient female mice are infertile due to anovulation [9], suggesting that P_4-PR signaling is essential for ovulation. This model is very useful for analyzing the physiological and pathological molecular pathways in ovulation, but not for the detailed evaluation of P_4's role in implantation and the subsequent pregnancy events. Nonetheless, we can utilize PR-null mice to understand the hormonal responsiveness of the uterus. Ovariectomized PR-null and wild-type (WT) mice treated with both E_2 and P_4 show different endometrial statuses of cell proliferation and differentiation [9]. It is widely accepted that cell proliferation is poorly compatible with differentiation, and distinct switching between proliferation and differentiation has been demonstrated for many different cell types [10–13]. In WT uteri, both attenuated proliferation of endometrial epithelial cells and activated proliferation of stromal cells start simultaneously [14]. Here, I describe this phenomenon as endometrial proliferation-differentiation switching (PDS). In contrast, PR-null uteri do not demonstrate PDS, but show epithelial proliferation and poor stromal cellularity [9]. These findings suggest that P_4-PR signaling leads to endometrial PDS. In the physiological condition when newly formed corpus luteum produces P_4 after ovulation in WT mice, P_4-PR signaling governs the uterus and induces endometrial PDS in the preimplantation period. In fact, an administration of a PR antagonist RU486 in the peri-implantation period impairs endometrial PDS and blastocyst implantation in WT mice [14]. In addition, endometrial PDS in the receptive uterus occurs not only in mice but also in humans [14]. Previous studies have definitely demonstrated that any types of genetically modified mice lacking endometrial PDS in the peri-implantation period do not have successful implantation outcome [1, 2, 4, 8, 15–17], strongly suggesting that endometrial PDS is a marker of uterine receptivity. Furthermore, PR has two isoforms, PR-A and PR-B, and previous reports have shown that PR-A is principally responsible for uterine function during pregnancy, contributing to endometrial PDS [18, 19]. However, it is assumed that PR-B does not have a critical function in pregnancy, because global ablation of PR-B does not show any issues in pregnancy outcome [18, 19]. Accordingly, the signaling of P_4-PR, especially of P_4-PR-A, controls endometrial PDS as well as uterine receptivity to embryo implantation.

2.5 Useful Models for Analyzing P_4 Signaling in Uterine Receptivity and Implantation

In addition to PR-deficient mice, there are several useful mouse models to clarify the roles of P_4 signaling in the uterus. Appropriate PR function depends on the stability of the PR complex. The functionally mature PR complex consists of a receptor monomer, a 90-kDa heat shock protein (Hsp90) dimer, the cochaperone, p23, and one of four cochaperones which include a tetratricopeptide repeat (TPR) that binds to Hsp90 [17, 20, 21]. The immunophilin cochaperone, FK506-binding protein 4 (FKBP52), is one of these TPR-containing chaperones, binding both Hsp90 and PR, stabilizing the structure of the PR complex, thereby reinforcing P_4-PR signaling [17, 20, 21]. Targeted deletion of FKBP52 attenuates uterine P_4-PR signaling, but does not completely suppress it, because minimal binding of P_4 to PR is retained [17, 20, 21]. Excessive P_4 administration can strengthen PR signaling in the uterus on a CD1 background, a notable characteristic of FKBP52-deficient mice, which is different from PR-null mice with loss of P_4 signaling [17]. Moreover, FKBP52-null females on the CD1 background show normal ovulation and normal P_4 secretion [17]. Therefore, unlike PR-null mice, the CD1 FKBP52-null mice are very useful tools for exploring the molecular mechanisms of P_4-PR signaling in the physiological processes of pregnancy after ovulation, including implantation, decidualization, and pregnancy maintenance. Previous investigations have demonstrated that FKBP52-deficient mice display diminished uterine responsiveness to P_4 and enhanced sensitivity to E_2, which disturbs the proper regulation of endometrial PDS in the preimplantation period, thus ultimately leading to implantation failure [20]. However, these disorders of endometrial PDS and embryo implantation in the CD1 FKBP52-deficient mice can be totally recovered by modest supplementation of P_4 via silastic implants of P_4 [17], indicating that P_4-PR signaling plays a crucial role in implantation. Consequently, FKBP52-deficient mouse is an established unique animal model reflecting what is known as "P_4 resistance," the diminished uterine responsiveness to P_4 which is reversed by P_4 supplementation in a genetic background-dependent manner.

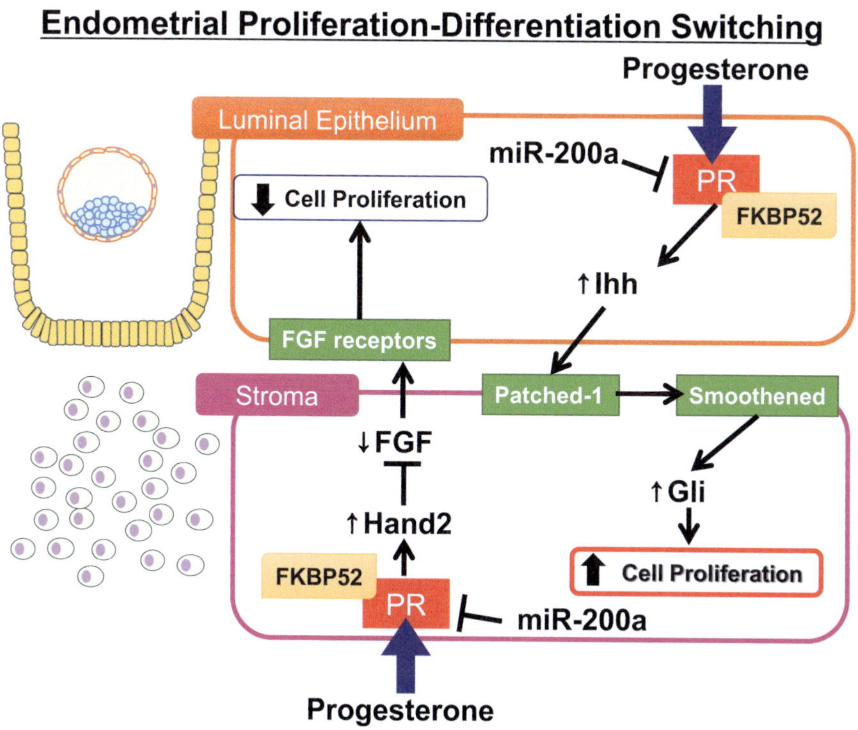

Fig. 2.4 Endometrial proliferation-differentiation switching in the receptive uterus

Other mouse models have also been performed to clarify the downstream targets of P_4-PR signaling. A microarray study of WT uteri with RU486 treatment during the preimplantation phase revealed that heart- and neural crest derivative-expressed protein 2 (Hand2), a basic helix-loop-helix transcription factor, is expressed in the endometrial stroma under the influence of P_4-PR signaling and inhibits epithelial cell proliferation through the downregulation of fibroblast growth factor (Fig. 2.4) [16]. Uterine deletion of Hand2 leads to implantation failure, confirming that it is essential for embryo attachment [16]. Another microarray study of PR-null uteri identified Indian hedgehog (Ihh), a hedgehog family molecule, as a downstream factor of PR, which is highly expressed in the uterine endometrial epithelium in WT mice just before implantation [22, 23]. Ihh functions via its receptor Patched-1 (Ptch1) which is locally expressed in the endometrial stroma and induces stromal proliferation, thus regulating the uterus for implantation (Fig. 2.4) [22, 23]. The proposed downstream targets of Ihh signaling are transcriptional factor Gli proteins and a nuclear receptor chicken ovalbumin upstream promoter-transcriptional factor (COUP-TFII). It has been suggested that Gli proteins contribute to stromal proliferation [22], and COUP-TFII modifies the balance between ER and PR signaling [24]. These findings show the presence of epithelial-stromal interactions under the control of ovarian hormones.

2.6 Regulation of P_4-PR Signaling by MicroRNA in Implantation

As described above, endometrial PDS, which is dependent on P_4-PR signaling, can be an index of uterine receptivity to the embryo [17, 20]. My research group recently discovered that endometrial PDS occurs in a spatial manner, between the uterus and cervix [14]. Under the normal conditions except the pathological conditions of ectopic pregnancy, blastocysts implant in the uterus, not the cervix. The endometrium in the mouse uterus shows PDS, while the cervix does exhibit any changes in proliferation or differentiation in neither the epithelium nor the stroma. Similarly in humans, the uterus presents dynamic PDS from the proliferative phase to the secretory phase, in contrast to the cervix which shows no significant changes in the cell proliferation status [14]. These findings suggest that mechanisms of regulation of P_4-PR signaling are different between the uterus and cervix and the diminished P_4-PR signaling prevent embryo implantation in the cervix. Consequently, comparing molecular signals between the uterus and cervix may help us to identify the important factors involving P_4-PR signaling during implantation. My group also found that P_4-PR signaling in the cervix is downregulated by microRNA (miR)-200a in two different pathways. First, elevated miR-200a expression in the cervix directly reduces PR protein levels by posttranscriptional regulation [14]. Then, miR-200a induces upregulation of 20α-hydroxysteroid dehydrogenase (20α-HSD), a P_4-metabolizing enzyme, through downregulation of Stat5, as previously reported [25], leading to the local metabolism of P_4 in the cervix. Moreover, miR-200a expression levels are low in the receptive uterus compared to the pre-receptive uterus (unpublished observation), indicating that the reduction of miR-200a induces the heightened uterine P_4-PR signaling which contributes to successful implantation (Fig. 2.4). These findings indicate that epigenetic regulation of P_4-PR signaling is involved in embryo implantation.

2.7 Interactions Between Signaling Pathways of E_2 and P_4 in Implantation

Appropriate balance between E_2 and P_4 signaling defines uterine receptivity in a very sophisticated manner. In mice, a small rise in ovarian E_2 secretion just before implantation with preceding ovarian P_4 production strictly rules the "implantation window," the time-limited acquisition of receptivity to embryo implantation in the uterus. Too much or too little E_2 results in opening defect of implantation window [26]. The linkage between pre-receptive ovarian E_2 secretion and uterine receptivity is controversial in primates [27–29]. However, in humans, the principle of the "implantation window" is generally accepted [30], and accumulated evidence has shown that heightened E_2 signaling disturbs the expression of essential molecules for implantation, such as integrin, leading to a higher rate of implantation failure

[31–34]. Implantation failure due to this aberrant hormonal signaling balance is also observed in knockout mouse models other than FKBP52-null mice. Uterine-specific deletion of the nuclear receptor co-activator 2 (Ncoa2) gene encoding steroid receptor co-activator 2 (SRC2) leads to implantation failure, inhibiting the optimization of the PR function by Ncoa2 [35]. Thus, Ncoa2 has a role in mediating P_4-PR signaling in the endometrium [35–37]. Although in vitro studies have reported that nuclear receptor co-activator 6 (Ncoa6) interacts with ERα as a co-activator [38–41], an in vivo study showed that Ncoa6 does not act as a co-activator but promotes the ubiquitination and degradation of ERα, attenuating uterine E_2-ER signaling [42]. Uterine ablation of Ncoa6 induces accumulation of ERα and enhances E_2 sensitivity, leading to the disruption of E_2/P_4 signaling balance and implantation failure [42]. Interestingly, not only this imbalance of hormonal signaling but also implantation failure is rescued by treatment with an ER antagonist ICI-182780 [42]. Mice with uterine depletion of the signal transducer and activator of transcription 3 (Stat3), known as a downstream molecule of leukemia inhibitory factor (LIF) before implantation [43], also show implantation failure with greater influence of E_2-ER than P_4-PR signaling on the uterus in the preimplantation period [44]. However, the detailed mechanism of Stat3 and E_2/P_4 signaling has not yet been fully elucidated.

Under the influences of P_4 and E_2, the endometrium secretes important mediators for cell-to-cell communications in the uterine microenvironments during implantation: cytokines and growth factors such as LIF and heparin-binding epidermal growth factor-like growth factor (HB-EGF) (Fig. 2.3). Maternal LIF is essential for successful implantation [45], and HB-EGF plays a key role in a two-way communication between the embryo and the uterus [46]. I describe these factors in the following sections.

2.8 Leukemia Inhibitory Factor (LIF)

LIF is a cytokine in interleukin-6 family members. LIF-deficient mice reveal complete implantation failure [45]. In addition, the phenotype of implantation failure is recovered by the administration of recombinant LIF protein to the knockout females [45, 47]. LIF-null embryos can develop normally and implant in WT uteri after blastocyst transfer to WT recipients; however, wild-type embryos do not implant in LIF-deficient uteri after blastocyst transfer to the null females [45, 48, 49]. These findings indicate that maternal LIF is critical for successful implantation.

LIF is an E_2-responsive gene in the mouse uterus, and LIF expression rapidly increases after E_2 injection in the uterus of ovariectomized mice [50, 51]. In fact, LIF is expressed at the highest level on day 1 of pregnancy when the uterus is under the influence of preovulatory E_2 surge. Thereafter, it is expressed in uterine glands on day 4 morning and then in the stroma surrounding the blastocyst at the time of the attachment reaction on day 4 night and persists through day 5 morning (Fig. 2.3)

[50]. Thus, LIF is expressed in day 4 pregnant uteri at two different times in two different cell types, and its expression is low in the post-implantation period [50]. These findings indicate that LIF is not required for pregnancy maintenance but for implantation. Nonetheless, precise effects of maternal LIF on implantation, especially on uterine receptivity and blastocyst activation, remain unclear.

In the study performed by my research group, mice with uterine deletion of p53 show normal implantation in spite of the reduction of LIF levels on day 4 morning [52]. Stromal LIF expression pattern surrounding the blastocyst at the time of attachment on day 4 midnight is normal in p53-null females [52], suggesting that eliminated LIF levels in p53-deleted uteri on day 4 morning is not a limiting factor for implantation. In addition, CD1 mice with deficiency of PR cochaperone FKBP52 show implantation failure due to P_4 resistance, and LIF expression of these mice is reduced at the glandular epithelium on day 4 morning and at the stroma on day 4 night [17]. P_4 supplementation to the mutant mice can reverse both the phenotype of defective implantation and stromal LIF expression on day 4 midnight, although LIF expression at the glandular epithelium on day 4 morning is still reduced after P_4 treatment [17]. These findings also suggest that stromal LIF on day 4 midnight may be more important than epithelial LIF on day 4 morning. Nonetheless, it is controversial where and when uterine LIF is expressed more critically on day 4 of pregnancy, and further investigations are required to clarify this issue (Fig. 2.3).

In association with LIF, MSX homeobox genes are reported to be essential transcriptional regulators which morphologically modulate luminal epithelium and control normal implantation in mice (Fig. 2.3) [15]. LIF reduces uterine Msx1 expression and deficiency of both Msx1 and Msx2 reduces LIF expression [15]. Importantly, uterine deletion of Msx1/2 completely inhibits blastocyst implantation [15]. These findings suggest that MSXs are one of the critical modulators in the system of uterine LIF expression in the peri-implantation period.

LIF binds LIF receptor which dimerizes with glycoprotein gp130, the common signaling receptor for IL-6 family cytokines, to activate several signaling pathways including JAK-STAT pathway, MAPK pathway, and PI3K-AKT pathway. In mouse embryonic stem (ES) cells, LIF upregulates Klf4 through JAK-STAT3 pathway and Tbx3 through PI3K-AKT pathway and strongly stimulates the expressions of Sox2 and Nanog to maintain the Oct3/4 expression [53]. In contrast, LIF also activates MAPK pathway to inhibit Tbx3 activity, suggesting that these downstream pathways of LIF coordinately regulate the differentiation of ES cells [53]. Compared with this, LIF does not activate MAPK but STAT3 in luminal epithelium on day 4 morning [43], suggesting the tissue-selective activation of signaling pathways by LIF.

2.9 Heparin-Binding Epidermal Growth Factor-Like Growth Factor (HB-EGF)

HB-EGF, one of EGF family members, is a key player in the embryo-uterine interactions with the subsequent uterine attachment reaction [46]. It is expressed in the luminal epithelium located around active blastocyst several hours before attachment [54]. HB-EGF is produced in soluble and transmembrane forms, and both forms affect embryonic functions in an autocrine, paracrine, and/or juxtacrine manner [54, 55] via the EGF family of receptors which is expressed on the cell surface of trophectoderm [56, 57]. The soluble form supports blastocyst growth [54], and the transmembrane form can make activated blastocysts adhere to the uterus [57] (Fig. 2.3). In addition, systemic deletion of HB-EGF leads to perinatal lethality [58], and its uterine deletion delays implantation and reduces litter size [58], emphasizing its importance in implantation.

HB-EGF and other EGF family members such as EGF, TGFα, betacellulin, epiregulin, neuregulin, and amphiregulin interact with the receptor subtypes of the ErbB family, ErbB1, ErbB2, ErbB3, and ErbB4, which have a tyrosine kinase domain for signal transduction. ErbBs form primarily homodimers or heterodimers to be activated by the ligands. Among these ErbB family members, ErbB1 and ErbB4 on the cell surface of trophectoderm can interact with uterine HB-EGF in implantation [56, 57]. The expression of both ErbB1 and ErbB4 is downregulated in dormant blastocyst but is markedly upregulated in the activated blastocyst [56, 59]. Activated blastocysts also express HB-EGF, which can induce uterine HB-EGF mRNA. These findings suggest the presence of a molecular feed-forward loop between the embryo and uterus for attachment reaction. Moreover, many studies also revealed significant roles of HB-EGF in human implantation. For instance, endometrial HB-EGF expression levels are highest in the receptive epithelium [60, 61]. The cells with the transmembrane form of HB-EGF can adhere to human blastocyst expressing cell surface ErbB4 [62]. Collectively, HB-EGF is critical for embryo-uterine interactions in implantation.

2.10 Conclusion

The number of babies born after the treatments using assisted reproductive technology is increasing along with the rise in the age of initial gestation and advances in techniques of in vitro fertilization [63]. In order to improve fertility rates, a lot of problems need to be solved, for instance, recurrent miscarriage despite the quality of transplanted embryos [64]. Implantation failure is one of the major causes of unexplained infertility, and also the most puzzling issue, since there are no effective treatments. Although previous mouse studies have revealed that uterine receptivity is regulated by key players such as ovarian hormones, cytokines, and growth factors, as described in this chapter, the detailed mechanisms are still unclear.

Further investigations are required to clarify them and to establish new strategies for implantation failure. New future findings are expected to be applied in a clinical setting for infertility treatment and contraception.

Acknowledgments This work was supported by JSPS KAKENHI Grant (Project Numbers: 24689062, 26670713, 26112506, 26112703, 40598653), the Cell Science Research Foundation, and GSK Japan Research Grant.

References

1. Egashira M, Hirota Y. Uterine receptivity and embryo-uterine interactions in embryo implantation: lessons from mice. Reprod Med Biol. 2013;12(4):127–32. doi:10.1007/s12522-013-0153-1.
2. Dey SK, Lim H, Das SK, Reese J, Paria BC, Daikoku T, et al. Molecular cues to implantation. Endocr Rev. 2004;25(3):341–73. doi:10.1210/er.2003-0020.
3. Urman B, Yakin K, Balaban B. Recurrent implantation failure in assisted reproduction: how to counsel and manage. A. General considerations and treatment options that may benefit the couple. Reprod Biomed Online. 2005;11(3):371–81.
4. Cha J, Sun X, Dey SK. Mechanisms of implantation: strategies for successful pregnancy. Nat Med. 2012;18(12):1754–67. doi:10.1038/nm.3012.
5. Paria BC, Lim H, Wang XN, Liehr J, Das SK, Dey SK. Coordination of differential effects of primary estrogen and catecholestrogen on two distinct targets mediates embryo implantation in the mouse. Endocrinology. 1998;139(12):5235–46. doi:10.1210/endo.139.12.6386.
6. Chaen T, Konno T, Egashira M, Bai R, Nomura N, Nomura S, et al. Estrogen-dependent uterine secretion of osteopontin activates blastocyst adhesion competence. PLoS One. 2012;7(11):e48933. doi:10.1371/journal.pone.0048933.
7. Hirota Y, Cha J, Dey SK. Revisiting reproduction: Prematurity and the puzzle of progesterone resistance. Nat Med. 2010;16(5):529–31. doi:10.1038/nm0510-529.
8. Wang H, Dey SK. Roadmap to embryo implantation: clues from mouse models. Nat Rev Genet. 2006;7(3):185–99. doi:10.1038/nrg1808.
9. Lydon JP, DeMayo FJ, Funk CR, Mani SK, Hughes AR, Montgomery Jr CA, et al. Mice lacking progesterone receptor exhibit pleiotropic reproductive abnormalities. Genes Dev. 1995;9(18):2266–78.
10. Conti L, Sipione S, Magrassi L, Bonfanti L, Rigamonti D, Pettirossi V, et al. Shc signaling in differentiating neural progenitor cells. Nat Neurosci. 2001;4(6):579–86. doi:10.1038/88395.
11. Dugan LL, Kim JS, Zhang Y, Bart RD, Sun Y, Holtzman DM, et al. Differential effects of cAMP in neurons and astrocytes. Role of B-raf. J Biol Chem. 1999;274(36):25842–8.
12. Chen JF, Mandel EM, Thomson JM, Wu Q, Callis TE, Hammond SM, et al. The role of microRNA-1 and microRNA-133 in skeletal muscle proliferation and differentiation. Nat Genet. 2006;38(2):228–33. doi:10.1038/ng1725.
13. Garcia AJ, Vega MD, Boettiger D. Modulation of cell proliferation and differentiation through substrate-dependent changes in fibronectin conformation. Mol Biol Cell. 1999;10(3):785–98.
14. Haraguchi H, Saito-Fujita T, Hirota Y, Egashira M, Matsumoto L, Matsuo M, et al. MicroRNA-200a locally attenuates progesterone signaling in the cervix, preventing embryo implantation. Mol Endocrinol. 2014;28(7):1108–17. doi:10.1210/me.2014-1097.
15. Daikoku T, Cha J, Sun X, Tranguch S, Xie H, Fujita T, et al. Conditional deletion of Msx homeobox genes in the uterus inhibits blastocyst implantation by altering uterine receptivity. Dev Cell. 2011;21(6):1014–25. doi:10.1016/j.devcel.2011.09.010.

16. Li Q, Kannan A, DeMayo FJ, Lydon JP, Cooke PS, Yamagishi H, et al. The antiproliferative action of progesterone in uterine epithelium is mediated by Hand2. Science. 2011;331 (6019):912–6. doi:10.1126/science.1197454.
17. Tranguch S, Wang H, Daikoku T, Xie H, Smith DF, Dey SK. FKBP52 deficiency-conferred uterine progesterone resistance is genetic background and pregnancy stage specific. J Clin Invest. 2007;117(7):1824–34. doi:10.1172/JCI31622.
18. Mulac-Jericevic B, Lydon JP, DeMayo FJ, Conneely OM. Defective mammary gland morphogenesis in mice lacking the progesterone receptor B isoform. Proc Natl Acad Sci U S A. 2003;100(17):9744–9. doi:10.1073/pnas.1732707100.
19. Mulac-Jericevic B, Mullinax RA, DeMayo FJ, Lydon JP, Conneely OM. Subgroup of reproductive functions of progesterone mediated by progesterone receptor-B isoform. Science. 2000;289(5485):1751–4.
20. Tranguch S, Cheung-Flynn J, Daikoku T, Prapapanich V, Cox MB, Xie H, et al. Cochaperone immunophilin FKBP52 is critical to uterine receptivity for embryo implantation. Proc Natl Acad Sci U S A. 2005;102(40):14326–31. doi:10.1073/pnas.0505775102.
21. Tranguch S, Smith DF, Dey SK. Progesterone receptor requires a co-chaperone for signalling in uterine biology and implantation. Reprod Biomed Online. 2007;14(Spec No 1):39–48. doi:10.1016/S1472-6483(10)61457-5.
22. Matsumoto H, Zhao X, Das SK, Hogan BL, Dey SK. Indian hedgehog as a progesterone-responsive factor mediating epithelial-mesenchymal interactions in the mouse uterus. Dev Biol. 2002;245(2):280–90. doi:10.1006/dbio.2002.0645.
23. Lee K, Jeong J, Kwak I, Yu CT, Lanske B, Soegiarto DW, et al. Indian hedgehog is a major mediator of progesterone signaling in the mouse uterus. Nat Genet. 2006;38(10):1204–9. doi:10.1038/ng1874.
24. Kurihara I, Lee DK, Petit FG, Jeong J, Lee K, Lydon JP, et al. COUP-TFII mediates progesterone regulation of uterine implantation by controlling ER activity. PLoS Genet. 2007;3(6):e102. doi:10.1371/journal.pgen.0030102.
25. Williams KC, Renthal NE, Condon JC, Gerard RD, Mendelson CR. MicroRNA-200a serves a key role in the decline of progesterone receptor function leading to term and preterm labor. Proc Natl Acad Sci U S A. 2012;109(19):7529–34. doi:10.1073/pnas.1200650109.
26. Ma WG, Song H, Das SK, Paria BC, Dey SK. Estrogen is a critical determinant that specifies the duration of the window of uterine receptivity for implantation. Proc Natl Acad Sci U S A. 2003;100(5):2963–8. doi:10.1073/pnas.0530162100.
27. Ghosh D, De P, Sengupta J. Luteal phase ovarian oestrogen is not essential for implantation and maintenance of pregnancy from surrogate embryo transfer in the rhesus monkey. Hum Reprod. 1994;9(4):629–37.
28. Smitz J, Bourgain C, Van Waesberghe L, Camus M, Devroey P, Van Steirteghem AC. A prospective randomized study on oestradiol valerate supplementation in addition to intravaginal micronized progesterone in buserelin and HMG induced superovulation. Hum Reprod. 1993;8(1):40–5.
29. Rao AJ, Ramachandra SG, Ramesh V, Krishnamurthy HN, Ravindranath N, Moudgal NR. Establishment of the need for oestrogen during implantation in non-human primates. Reprod Biomed Online. 2007;14(5):563–71.
30. Wilcox AJ, Baird DD, Weinberg CR. Time of implantation of the conceptus and loss of pregnancy. N Engl J Med. 1999;340(23):1796–9. doi:10.1056/NEJM199906103402304.
31. Diana M, Schettini M, Gallucci M. Evaluation and management of malfunctionings following implantation of the artificial urinary sphincter. Int Surg. 1999;84(3):241–5.
32. Gregory CW, Wilson EM, Apparao KB, Lininger RA, Meyer WR, Kowalik A, et al. Steroid receptor coactivator expression throughout the menstrual cycle in normal and abnormal endometrium. J Clin Endocrinol Metab. 2002;87(6):2960–6. doi:10.1210/jcem.87.6.8572.
33. Apparao KB, Lovely LP, Gui Y, Lininger RA, Lessey BA. Elevated endometrial androgen receptor expression in women with polycystic ovarian syndrome. Biol Reprod. 2002;66 (2):297–304.

34. Khorram O, Lessey BA. Alterations in expression of endometrial endothelial nitric oxide synthase and alpha(v)beta(3) integrin in women with endometriosis. Fertil Steril. 2002;78 (4):860–4.
35. Mukherjee A, Amato P, Allred DC, DeMayo FJ, Lydon JP. Steroid receptor coactivator 2 is required for female fertility and mammary morphogenesis: insights from the mouse, relevance to the human. Nucl Recept Signal. 2007;5:e011. doi:10.1621/nrs.05011.
36. Xu J, Wu RC, O'Malley BW. Normal and cancer-related functions of the p160 steroid receptor co-activator (SRC) family. Nat Rev Cancer. 2009;9(9):615–30. doi:10.1038/nrc2695.
37. Mukherjee A, Soyal SM, Fernandez-Valdivia R, Gehin M, Chambon P, Demayo FJ, et al. Steroid receptor coactivator 2 is critical for progesterone-dependent uterine function and mammary morphogenesis in the mouse. Mol Cell Biol. 2006;26(17):6571–83. doi:10.1128/MCB.00654-06.
38. Mahajan MA, Samuels HH. A new family of nuclear receptor coregulators that integrate nuclear receptor signaling through CREB-binding protein. Mol Cell Biol. 2000;20 (14):5048–63.
39. Lee SK, Anzick SL, Choi JE, Bubendorf L, Guan XY, Jung YK, et al. A nuclear factor, ASC-2, as a cancer-amplified transcriptional coactivator essential for ligand-dependent transactivation by nuclear receptors in vivo. J Biol Chem. 1999;274(48):34283–93.
40. Ko L, Cardona GR, Chin WW. Thyroid hormone receptor-binding protein, an LXXLL motif-containing protein, functions as a general coactivator. Proc Natl Acad Sci U S A. 2000;97 (11):6212–7.
41. Caira F, Antonson P, Pelto-Huikko M, Treuter E, Gustafsson JA. Cloning and characterization of RAP250, a novel nuclear receptor coactivator. J Biol Chem. 2000;275(8):5308–17.
42. Kawagoe J, Li Q, Mussi P, Liao L, Lydon JP, DeMayo FJ, et al. Nuclear receptor coactivator-6 attenuates uterine estrogen sensitivity to permit embryo implantation. Dev Cell. 2012;23 (4):858–65. doi:10.1016/j.devcel.2012.09.002.
43. Cheng JG, Chen JR, Hernandez L, Alvord WG, Stewart CL. Dual control of LIF expression and LIF receptor function regulate Stat3 activation at the onset of uterine receptivity and embryo implantation. Proc Natl Acad Sci U S A. 2001;98(15):8680–5. doi:10.1073/pnas.151180898.
44. Sun X, Bartos A, Whitsett JA, Dey SK. Uterine deletion of Gp130 or Stat3 shows implantation failure with increased estrogenic responses. Mol Endocrinol. 2013;27(9):1492–501. doi:10.1210/me.2013-1086.
45. Stewart CL, Kaspar P, Brunet LJ, Bhatt H, Gadi I, Kontgen F, et al. Blastocyst implantation depends on maternal expression of leukaemia inhibitory factor. Nature. 1992;359(6390):76–9. doi:10.1038/359076a0.
46. Lim HJ, Dey SK. HB-EGF: a unique mediator of embryo-uterine interactions during implantation. Exp Cell Res. 2009;315(4):619–26. doi:10.1016/j.yexcr.2008.07.025.
47. Cheng JG, Rodriguez CI, Stewart CL. Control of uterine receptivity and embryo implantation by steroid hormone regulation of LIF production and LIF receptor activity: towards a molecular understanding of "the window of implantation". Rev Endocr Metab Disord. 2002;3 (2):119–26.
48. Chen JR, Cheng JG, Shatzer T, Sewell L, Hernandez L, Stewart CL. Leukemia inhibitory factor can substitute for nidatory estrogen and is essential to inducing a receptive uterus for implantation but is not essential for subsequent embryogenesis. Endocrinology. 2000;141 (12):4365–72.
49. Sherwin JR, Freeman TC, Stephens RJ, Kimber S, Smith AG, Chambers I, et al. Identification of genes regulated by leukemia-inhibitory factor in the mouse uterus at the time of implantation. Mol Endocrinol. 2004;18(9):2185–95. doi:10.1210/me.2004-0110.
50. Song H, Lim H, Das SK, Paria BC, Dey SK. Dysregulation of EGF family of growth factors and COX-2 in the uterus during the preattachment and attachment reactions of the blastocyst with the luminal epithelium correlates with implantation failure in LIF-deficient mice. Mol Endocrinol. 2000;14(8):1147–61.

51. Bhatt H, Brunet LJ, Stewart CL. Uterine expression of leukemia inhibitory factor coincides with the onset of blastocyst implantation. Proc Natl Acad Sci U S A. 1991;88(24):11408–12.
52. Hirota Y, Daikoku T, Tranguch S, Xie H, Bradshaw HB, Dey SK. Uterine-specific p53 deficiency confers premature uterine senescence and promotes preterm birth in mice. J Clin Invest. 2010;120(3):803–15. doi:10.1172/JCI40051.
53. Niwa H, Ogawa K, Shimosato D, Adachi K. A parallel circuit of LIF signalling pathways maintains pluripotency of mouse ES cells. Nature. 2009;460(7251):118–22. doi:10.1038/nature08113.
54. Das SK, Wang XN, Paria BC, Damm D, Abraham JA, Klagsbrun M, et al. Heparin-binding EGF-like growth factor gene is induced in the mouse uterus temporally by the blastocyst solely at the site of its apposition: a possible ligand for interaction with blastocyst EGF-receptor in implantation. Development. 1994;120(5):1071–83.
55. Das SK, Tsukamura H, Paria BC, Andrews GK, Dey SK. Differential expression of epidermal growth factor receptor (EGF-R) gene and regulation of EGF-R bioactivity by progesterone and estrogen in the adult mouse uterus. Endocrinology. 1994;134(2):971–81.
56. Paria BC, Elenius K, Klagsbrun M, Dey SK. Heparin-binding EGF-like growth factor interacts with mouse blastocysts independently of ErbB1: a possible role for heparan sulfate proteoglycans and ErbB4 in blastocyst implantation. Development. 1999;126(9):1997–2005.
57. Raab G, Kover K, Paria BC, Dey SK, Ezzell RM, Klagsbrun M. Mouse preimplantation blastocysts adhere to cells expressing the transmembrane form of heparin-binding EGF-like growth factor. Development. 1996;122(2):637–45.
58. Xie H, Wang H, Tranguch S, Iwamoto R, Mekada E, Demayo FJ, et al. Maternal heparin-binding-EGF deficiency limits pregnancy success in mice. Proc Natl Acad Sci U S A. 2007;104(46):18315–20. doi:10.1073/pnas.0707909104.
59. Paria BC, Das SK, Andrews GK, Dey SK. Expression of the epidermal growth factor receptor gene is regulated in mouse blastocysts during delayed implantation. Proc Natl Acad Sci U S A. 1993;90(1):55–9.
60. Yoo HJ, Barlow DH, Mardon HJ. Temporal and spatial regulation of expression of heparin-binding epidermal growth factor-like growth factor in the human endometrium: a possible role in blastocyst implantation. Dev Genet. 1997;21(1):102–8. doi:10.1002/(SICI)1520-6408(1997)21:1<102::AID-DVG12>3.0.CO;2-C.
61. Leach RE, Khalifa R, Ramirez ND, Das SK, Wang J, Dey SK, et al. Multiple roles for heparin-binding epidermal growth factor-like growth factor are suggested by its cell-specific expression during the human endometrial cycle and early placentation. J Clin Endocrinol Metab. 1999;84(9):3355–63.
62. Chobotova K, Spyropoulou I, Carver J, Manek S, Heath JK, Gullick WJ, et al. Heparin-binding epidermal growth factor and its receptor ErbB4 mediate implantation of the human blastocyst. Mech Dev. 2002;119(2):137–44.
63. van Loendersloot L, Repping S, Bossuyt PM, van der Veen F, van Wely M. Prediction models in in vitro fertilization; where are we? A mini review. J Adv Res. 2014;5(3):295–301. doi:10.1016/j.jare.2013.05.002.
64. Polanski LT, Baumgarten MN, Quenby S, Brosens J, Campbell BK, Raine-Fenning NJ. What exactly do we mean by 'recurrent implantation failure'? A systematic review and opinion. Reprod Biomed Online. 2014;28(4):409–23. doi:10.1016/j.rbmo.2013.12.006.

Chapter 3
Assessing Receptivity of the Human Endometrium to Improve Outcomes of Fertility Treatment

Tracey J. Edgell, Jemma Evans, Luk J.R. Rombauts, Beverley J. Vollenhoven, and Lois A. Salamonsen

Abstract Despite considerable improvements in assessment of embryo quality in infertility clinics, outcomes in terms of take-home baby rate have not improved substantially. Failure of the endometrium to achieve receptivity and the timing of the receptive period are now recognised as important issues in the success of IVF. Indeed, immunohistochemical and morphological studies show that the endometrium is highly disturbed in any cycle in which ovulation induction is performed, leading to recommendations that all embryos be frozen and replaced in a non-stimulation cycle. Assessment of any woman for endometrial receptivity either in cycles prior to treatment or testing for the potential for an embryo to implant in the cycle of transfer is urgently needed. However, careful consideration must be given to issues such as sampling, timing and rapid delivery of results as well as the best biomarkers, to enable in-clinic decision-making. Here these issues are

T.J. Edgell • J. Evans
Centre for Reproductive Health, Hudson Institute of Medical Research, 27-31 Wright St, Clayton, VIC 3168, Australia

L.J.R. Rombauts
Department of Obstetrics and Gynaecology, Monash University, Melbourne, VIC 3168, Australia

Monash IVF, Melbourne, VIC 3168, Australia

B.J. Vollenhoven
Department of Obstetrics and Gynaecology, Monash University, Melbourne, VIC 3168, Australia

Women's and Children's Programme, Monash Health, Melbourne, VIC 3168, Australia

Monash IVF, Melbourne, VIC 3168, Australia

L.A. Salamonsen (✉)
Centre for Reproductive Health, Hudson Institute of Medical Research, 27-31 Wright St, Clayton, VIC 3168, Australia

Department of Obstetrics and Gynaecology, Monash University, Melbourne, VIC 3168, Australia
e-mail: lois.salamonsen@hudson.org.au

H. Kanzaki (ed.), *Uterine Endometrial Function*,
DOI 10.1007/978-4-431-55972-6_3

considered, along with how endometrial receptivity testing might best be performed to optimise outcomes of infertility treatment.

Keywords Endometrial receptivity • Biomarkers • IVF success

3.1 Endometrial Receptivity

An absolute need for synchrony of development between the maternal endometrium and a developing blastocyst was clearly demonstrated in embryo transfer experiments in animals some decades ago [1, 2]. In these studies using sheep and rabbits, pregnancy could not be established if endometrial-embryo asynchrony was >3 days. Subsequently, Psychoyos coined the phrase 'window of implantation' for the period of time when the endometrium is optimally prepared for implantation [3]: this is now generally referred to as the phase of endometrial receptivity.

In women, it was noted as early as the 1950s that in uteri removed at hysterectomy from young women, embryos were found attached to the uterine wall only if the endometrium was in the mid-secretory phase of the menstrual cycle [4]. This first established that the receptive phase in humans occurs between 6 and 10 days after ovulation [5] and was confirmed in a now classic large epidemiological study in which 189 human conceptions were confirmed by the detection of human chorionic gonadotrophin (hCG) in maternal urine only 6–12 days after ovulation [6]. This study also defined the limits of the optimal phase of endometrial receptivity, demonstrating that implantation was possible after day 10 post-ovulation but was suboptimal: up to 82 % of pregnancies which occurred 11 days after ovulation resulted in miscarriage.

Attainment of endometrial receptivity is driven by progesterone acting on an oestrogen-primed endometrium. Gene array studies [7] indicate that endometrial gene regulation alters with time of progesterone exposure with early and late response genes. Importantly, progesterone mediates both increased and decreased expression of specific genes in human endometrial cells with decreased genes being in the majority [7]. While a number of differential gene array analyses have compared times of the cycle (mostly mid-proliferative versus mid-secretory), very few changes in specific genes have been common to all studies. These observations indicate the difficulty of comparing gene array data between laboratories using (a) different arrays, (b) different patient cohorts, (c) different sampling techniques and (d) different methods of data analysis. A complication in comparing such studies likely lies in the staging of endometrium used for analysis; the classic 'Noyes' criteria of dating the endometrium are now well accepted as inadequate [8], and a better dating technique based on molecular markers is needed. This may be partially fulfilled by the use of the newly developed endometrial receptivity array (ERA) (see below) although other measures of endometrial dating, and particularly determination of endometrial receptivity, are needed.

Key features of receptive endometrium cannot be identified by morphology alone, except when the endometrium is clearly inadequately developed for the expected phase of the menstrual cycle (such as lack of secretory features of glands and spiral arteriole development). Furthermore, analysis of entire endometrial biopsies by gene array or multiplex protein assays is confounded by the considerable variability between cellular compositions of any sample. These differ considerably in the proportions of epithelial cells to stromal fibroblasts, cells of the vasculature and leukocytes. Laser capture and gene array analysis of individual cellular compartments [9] has clearly demonstrated this issue. Additionally, as demonstrated by histological assessment of stimulated (IVF) endometria, different cellular compartments (e.g. epithelium and stroma) may be 'out of phase' with each other adding to the complication of such whole tissue analyses [9].

3.2 Implantation-Specific Microenvironments

The peri-implantation microenvironment within the uterine cavity is represented in Fig. 3.1. The embryo first enters the uterine cavity as an unhatched blastocyst and undergoes its final development through hatching to attachment to the uterine luminal epithelium within the environment of uterine fluid. This fluid contains a plethora of proteins, lipids, ions, amino acids, nutrients and microvesicles/ exosomes. The soluble factors can be derived from a number of sources including

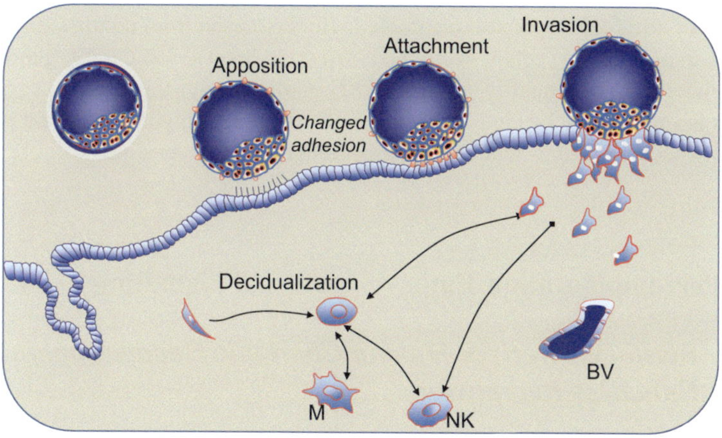

Fig. 3.1 Human embryo implantation. The embryo enters the uterine cavity as an unhatched blastocyst. After hatching it becomes apposed to the endometrial epithelium: the cell surfaces of both the trophectoderm and the endometrial luminal epithelium must change their adhesive properties to enable attachment and subsequent invasion of trophoblast cells into the decidualising stromal compartment, eventually to reach the blood vessels (BV). Macrophages (M) and uterine NK (uNK) cell numbers increase as decidualisation proceeds (Reproduced with permission from [10], CSIRO Publishing, http://www.publish.csiro.au/nid/44/paper/RD09145.htm)

selective transudation from the blood, carriage from the Fallopian tubes and likely also the peritoneal cavity and, importantly, secretions from the endometrial glands. Thus, it is likely that factors released or accumulated during the mid-secretory phase, when implantation takes place, and also endometrial secretions from the early secretory phase will be important for the final stages of blastocyst development.

The next important milieu for implantation is that of the developing decidua, which the blastocyst encounters once it has traversed the luminal epithelium. In women, differentiation of the stromal fibroblasts to decidual cells also occurs within each cycle, regardless of whether or not conception has occurred. Decidual development is dependent on progesterone acting through its receptors on the stromal cells: however, once initiated, a plethora of cytokines and other factors are released which, in turn, act on adjacent stromal cells in a cascade of intracellular events leading to a much wider decidualisation. These factors which include interleukin 11, activin A and relaxin act through separate pathways including but not restricted to cAMP [10, 11]. Decidualised stromal cells also secrete chemokines which act as chemoattractants for the macrophages and uterine natural killer (uNK) cells that are essential components of the decidua of pregnancy. The decidual milieu is overall favourable for trophoblast expansion and migration [12, 13]. Recently, early decidua has been identified as a 'sensor' of human embryo quality [14, 15]. It was found that the decidual cells could discriminate between 'good' and 'bad' quality embryos, except in the case of cells derived from women with recurrent pregnancy loss in whom the discriminatory capacity was absent [16]. Additionally, arrested 'bad-quality' embryos inhibited secretion of pro-implantation factors by decidualising stromal cells from normal women [15]. Genome-wide expression profiling of decidual responses to soluble factors released from competent embryos showed that only 15 genes were responsive, whereas some 449 genes were dysregulated by poor quality embryos [12]. Collectively, these data suggest the decidua is responsible for determining whether or not a pregnancy should proceed following successful pre- and early implantation events.

3.3 Peri-implantation Embryo-Maternal Signalling

3.3.1 Blastocyst-Derived Human Chorionic Gonadotrophin Enhances Receptivity

Human chorionic gonadotrophin (hCG) is the most well-known product of the developing embryo. Its gene expression is detectable by the time the blastocyst is formed [17] and hCG levels in serum underpin early pregnancy detection kits. hCG is formed of an alpha- and a beta-subunit and has considerable similarity to luteinising hormone (LH) with which it shares a common subunit. It binds to both the LH/hCG receptor and a mannose receptor [18] through which its signals are transduced: LH/hCG receptor mRNA [19] and protein [20] are present in the endometrium; immunohistochemistry shows them located to mid-secretory phase luminal epithelium. hCG exists naturally in minimally glycosylated and hyper-glycosylated forms, a sulphated form and as a free beta-subunit which potentially have different bioactivities. The highly glycosylated form is produced in trophoblast-derived cells of the placenta [13]. Since acidic forms of hCG are secreted by early blastocysts, it is likely that these are likewise highly glycosylated [21]. An effect of hCG on the endometrium was clearly demonstrated in an elegant in vivo study in which hCG was infused into the uterine cavity of women in the mid-secretory phase and found to induce production of pro-implantation factors: leukaemia inhibitory factor (LIF) and vascular endothelial growth factor (VEGF) [22], observations reinforced by studies in non-human primates [23]. In vitro, hCG stimulates secretion of selected cytokines by endometrial epithelial cells, confirming LIF and VEGF as hCG targets but also identifying IL-11, FGF2, GM-CSF and CXCL10 and prokineticin 1 as novel hCG-induced factors [24, 25]. Since all of these have known pro-implantation functions, it is clear that during a conception cycle, blastocyst-derived hCG acts to enhance endometrial receptivity. Given the progress in proteomic technologies, it is likely that other secreted human blastocyst proteins of lower abundance will soon be identified and their functions elucidated.

3.3.2 Endometrial-Secreted Products Support Pre-implantation Blastocyst Development

Endometrial products, particularly proteins, secreted into the uterine lumen from the early to mid-secretory phases of the cycle can enhance features of blastocyst development in vitro and most likely promote both survival and development of blastocysts. While some data has suggested that adding individual growth factors and cytokines (including HB-EGF, IGF-1, LIF and GM-CSF) to blastocyst culture prior to embryo implantation can improve blastocyst development in vitro [26, 27], the only one of these followed through to clinical trials is GM-CSF. This had a modest positive effect on ongoing pregnancy rate and live birth rate [28], but only when the conventional level of HSA in the embryo culture medium was reduced. Given that uterine fluid contains multiple factors, it is important that their impact on blastocyst development is determined in combination. Exposure of human or mouse embryos to human uterine lavage in vitro has effects that differ depending upon the time of the cycle at which the lavage is harvested. Interestingly, mid-proliferative phase lavage is detrimental to embryo development: mouse blastocyst outgrowth on fibronectin is strongly and significantly inhibited. In contrast, pooled uterine lavage from the mid-secretory phase significantly enhances blastocyst outgrowth. In one study, this could be replicated by recombinant (r)VEGFA [29]. Furthermore, when mouse embryos were pretreated with either VEGF121, VEGF165 or rVEGFA, the time to cavitation and blastocyst number were also increased, and following transfer of these blastocysts to recipient mothers, both implantation rate and foetal limb development were enhanced [30]. Therefore, VEGF could be an important additive for embryo culture prior to IVF. However, in a study in mice, there was only a trend for VEFG165-treated and transferred embryos to improve viable pregnancies [30].

3.4 Lack of Synchrony in Controlled Ovarian Stimulation Cycles

3.4.1 Why Are IVF Success Rates So Low?

Nearly four decades after the birth of Louise Brown, the first baby to result from application of what is now commonly known as IVF treatment, the per-cycle success rates have not substantially increased worldwide, remaining at <30 % [31, 32]. This is in spite of technical advancement in embryo culture, selection and transfer. Indeed, implantation failure was identified as the most common outcome following embryo transfer [33]. The vital but often overlooked factor is the ‘soil for the seed’, the endometrium, which becomes receptive for embryo implantation only in the mid-secretory phase of the menstrual cycle in synchrony with blastocyst development. Inability to establish receptivity leads to infertility and is a major cause for implantation failure in IVF cycles. In many IVF cycles, the embryo is transferred without establishing whether the endometrium is likely to be receptive (Fig. 3.2). This is despite the woman undergoing considerable hormonal treatment to induce oocyte development and ovulation, which may impact on endometrial development, and a wealth of recent knowledge regarding molecular changes essential for receptivity.

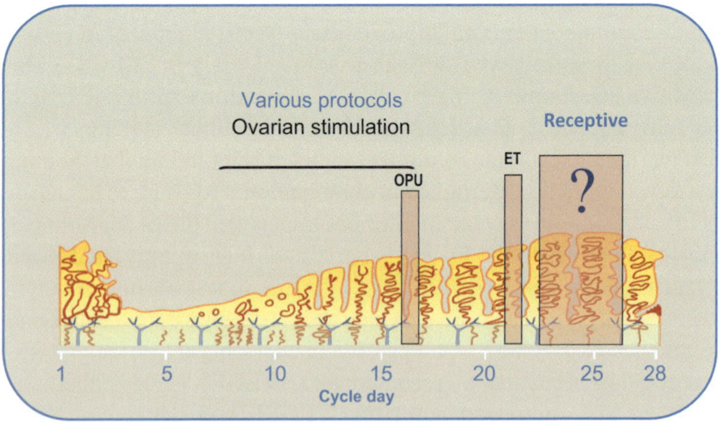

Fig. 3.2 Sequence of events in an IVF cycle. *OPU* day of ovum pickup, *E* day of embryo transfer

3.4.2 The Endometrium in Treatment Cycles

IVF cycles have severely abnormal hormone concentrations. This is due to the administration of gonadotrophins, as well as gonadotrophin-releasing hormone agonists and antagonists administered by differing protocols, resulting in very high oestrogen and progesterone concentrations, and compounded by the hCG administered for final oocyte maturation prior to oocyte harvest. Indeed, the precocious progesterone elevation often seen even as early as the day of hCG administration (0.08 ng/mL) is associated with a reduced probability of clinical pregnancy in fresh embryo transfer cycles, but not when the embryo is frozen and transferred in a normal cycle or in a donor-recipient cycle [34]. In addition, differential genomic analysis of endometrium from women with such elevated progesterone concentrations compared to women with normal progesterone concentrations revealed alterations in both miRNAs and mRNAs resulting from high progesterone exposure [35, 36]. However, while the endometrial receptivity profile (measured by DNA microarrays) in patients with premature progesterone elevation on the day of hCG administration showed an altered gene expression shift between the pre-receptive and receptive phases, it had no significant effect on a limited number of specific markers of endometrial receptivity [37] supporting earlier observations [38, 39]. These data, however, must be interpreted with caution as the gene arrays and subsequent validation studies were performed using a very small number of samples ($n = 4$ different women per group), and it is therefore important that such observations be tested further before applicability to all progesterone-elevated cycles can be determined. It is also important to remember that endometrial progesterone receptor (PR) expression changes across the cycle in response to alterations in oestrogen and progesterone concentrations. Specifically, epithelial PRA and PRB are lost during the secretory phase of the menstrual cycle with these receptors maintained in the stromal compartment. Epithelial responses to progesterone in the secretory phase are therefore mediated by local stromal factors via an indirect rather than a direct mechanism. Haouzi and colleagues [37] propose that premature progesterone elevation may lead to a precocious downregulation of epithelial progesterone receptor, prematurely inhibiting the endometrial response to this hormone. Such cell compartment-specific responses may be masked in gene array analysis which examines whole tissue, and 'compartment-specific' analysis may be more appropriate for future studies [9]. Nevertheless, in accord with the historical animal studies, there is a complete failure to achieve implantation during IVF/ICSI cycles with stimulation by either gonadotrophin-releasing hormone (GnRH) agonist or antagonist, when histological dating (according to Noyes' criteria) [40] shows dys-synchrony of >3 days [41–44]. Further, microarray studies have indicated that ovarian stimulation may alter the receptive phase endometrium such that it is detrimental to implantation [45]. Probably the strongest proof to date comes from a recent comprehensive immunohistochemical and histological study of tissues taken on LH/hCG+2 (at the time of oocyte pickup) from normal women, and women stimulated with either agonist or antagonist protocols and hCG, emphasising that the disturbance of the endometrium is much more than just developmental advancement [46]. The parameters investigated immunohistochemically included the progesterone receptor, leukocytes (CD45) and their subsets (uNK cells, CD56; macrophages, CD68; activated neutrophils, elastase), decidualised stromal cells (prolactin) and vasculature (CD34), in addition to morphological features (glandular development, oedema, blood vessel size). All parameters were scored and normalised against those for normal cycling women on LH +2. In the agonist stimulation group, outcomes of embryo transfer (pregnant or not pregnant) enabled stratification of data according to pregnancy outcome. Key data is summarised in Fig. 3.3 and clearly indicates that ovarian stimulation severely influences endometrial development. Of prime importance is that the cohort of women who did become pregnant following embryo transfer showed significantly less endometrial disturbance than those who did not become pregnant. The endometrium of women who failed to become pregnant also contained highly activated neutrophils (Fig. 3.3a), a state which is normally only seen at menstruation where they contribute strongly to tissue breakdown and repair. No doubt these contribute to the 'menstrual-like features' in some of the tissues. In addition, this data, showing extreme disturbance to the endometrium as early as hCG+2 in a

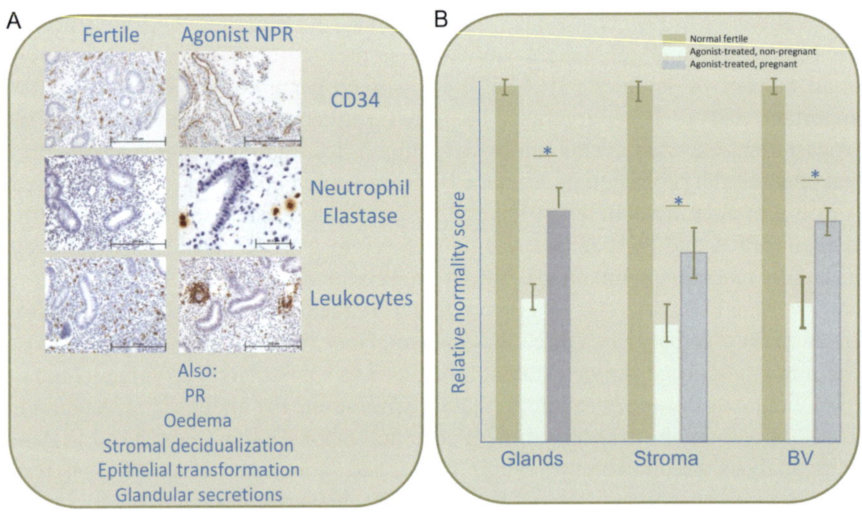

Fig. 3.3 Histological and immunohistochemical analysis was performed on endometrial biopsies taken 2 days after ovulation induction (OI+2) and on biopsies from normal cycling women on LH +2. Women in the OI group were stimulated by an agonist protocol and retrospectively separated into groups of women who did become pregnant or did not become pregnant (Pr) following fresh embryo transfer. Nine parameters as listed were examined. (**a**) Immunohistochemistry identifying blood vessels (CD34) and neutrophil activation (extracellular elastase). (**b**) Combined data from analysis of all parameters showing % of normal features (normal tissue expressed as 100 %) in glands, stroma and blood vessels. Note the high level of disturbance (<50 % normal) in all stimulated cycles and that this is significantly less in the women who became pregnant compared with those who did not become pregnant. *ER* oestrogen receptor, *PR* progesterone receptor (Derived from data in [46])

stimulation cycle, indicates that it may be possible to predict as early as the day of ovum pickup whether, based on the likelihood of not developing a receptive endometrium, the embryo should be frozen or whether it has a strong probability of implanting following fresh embryo transfer.

A detrimental effect on endometrial receptivity of the hCG administered for final oocyte maturation in ovulation induction as part of an IVF cycle has also been demonstrated. As detailed above, blastocyst hCG enhances endometrial receptivity. However, in women undergoing IVF cycles, the hCG receptor, by which the effects of hCG are transduced, is dramatically downregulated in the endometrial epithelium. Furthermore, functional studies in vitro show that the response to acute administration of hCG (mimicking blastocyst-secreted hCG), in terms of both endometrial adhesiveness and tight junction integrity, is lost if the acute hCG is preceded by a low chronic treatment of hCG (mimicking hCG administration during an IVF cycle) [20]. Thus, in an IVF stimulation cycle, the natural responsiveness of the endometrium to blastocyst hCG is lost, decreasing the likelihood of successful implantation.

These data, in concert with a number of clinical studies [47, 48], strongly support frozen embryo transfer rather than the fresh embryo transfer most commonly used. This would overcome the detrimental effects of ovarian stimulation and induced ovulation.

3.5 Testing for Receptivity

Given that endometrial receptivity is essential to establish pregnancy, testing for receptivity or, ideally, for prediction of receptivity later in the same cycle may provide a major boost for IVF success rates per cycle. However, consideration must be given to which form of testing would be most useful.

3.5.1 What to Measure

Many potential individual biomarkers for receptivity have been identified within research programmes studying mechanisms of implantation (Table 3.1). In some studies the change between proliferative and mid-secretory phase of fertile women has been examined essentially to identify critical components of receptivity. Other studies have directly compared mid-secretory (based on Noyes criteria) samples from fertile and infertile women with idiopathic infertility. In most cases, comparison of their expression in infertile versus fertile women has not had the power for analysis of specificity and sensitivity. The newer technologies available for biomarker discovery utilise 'omics' technologies [49] that include genomics, transcriptomics, epigenomics, lipidomics, proteomics and metabolomics. The advantage of these methods is that they can differentially identify a vast number

Table 3.1 Examples of individual proteins validated as differentiating between receptive and non-receptive endometrium

Protein	Sample assayed	References
Proprotein convertase 5/6	Uterine fluid	[83]
Integrin β3	Tissue biopsy	[84]
Vascular endothelial growth factor A	Uterine fluid	[79]
Stathmin 1	Tissue biopsy	[85]
Annexin A2	Tissue biopsy	[85]
α-Dystroglycan-N fragment	Uterine fluid	[74]
Progesterone receptor membrane component 1	Tissue biopsy	[86]
Glycodelin A	Uterine fluid	[87, 88]
α2-Macroglobulin	Uterine fluid	[79]
Antithrombin III	Uterine fluid	[79]

of molecular changes in matched samples: the disadvantage is they have generated large lists of 'potential' markers, for which validation has been limited or non-existent. Genomics, transcriptomics and epigenomics require tissue samples, whereas proteomics, lipidomics and metabolomics can be applied also to body fluids and are thus less invasive. Appropriate handling of samples prior to analysis is essential: RNA, proteins and many lipids are readily degraded. These limitations have been described in detail [49]. These powerful techniques themselves continue to evolve and gain increased power and are revolutionising the search for biomarkers.

Simultaneous measurement of more than one marker is becoming the standard for biomarker assessment in most fields, including cancers. Given the molecular complexity of disorders, it is unlikely that any one specific marker will discriminate or diagnose pathology. Particularly with regard to endometrial receptivity, defects probably represent a range of molecular changes, making it imperative that a large panel of biomarkers is used. Such multiple analysis or 'multiplexing' approach is well served by the 'omics' techniques. Given that many individual biomarkers of receptivity have been proposed, there is now a critical need to test these in combination to determine their collective power to differentiate between receptive and non-receptive endometrium.

3.5.2 *Type of Sample*

A number of options are available for sampling. Tissue sampling, usually by pipelle biopsy, is the most invasive [50], particularly for nulliparous women, but has been used in most tests available to date. Uterine aspiration and lavage are much less invasive. Given the very small volume of uterine fluid (<10 μl is retrieved by aspiration) and that aspirates usually contain blood indicating damage to the tissue, our laboratory and others favour a uterine lavage with 2–3 mL saline, gently infused into the uterine cavity and retrieved so that it washes over the entire endometrial surface. Importantly, aspiration and lavage are not interchangeable for the purpose of analyte analysis [51], presumably because soluble analytes are released from the endometrial glycocalyx during lavage. Aspiration may be the better technique if sampling is to be performed in the same cycle as embryo transfer, as it does not appear to influence implantation rates [52]. However, as indicated above, collection of an aspirate does carry a risk of tissue damage within the uterus and may therefore compromise the endometrium at the time of implantation. Uterine lavage offers a greater range of proteins for assessment of multiple factors, thereby increasing the sensitivity and specificity of the predictive test. Clearly a minimally invasive test based on blood, urine or saliva would be optimal, as testing need not be limited to days when the patient is in the clinic, and indeed, consecutive days of testing to identify the optimal transfer time then becomes possible. However, this presents a challenge since most of the factors identified as potential biomarkers are produced locally by the endometrium in very low concentrations and not secreted directly into any of these fluids. Quality of the sample is of utmost importance: both collection and storage require adherence to strict standard operating procedures.

3.5.3 *Variability in Timing of Receptivity Within and Between Women*

The highly dynamic nature of the endometrium makes obtaining clinical samples extremely difficult. Uniquely, the cellular and molecular composition of the endometrium alters on a daily basis making histological dating of the endometrium by Noyes' classification highly variable, with considerable observer error [8]. Molecular markers are required to provide objective dating and prediction of receptivity. The recently described endometrial array, the ERA [53], suggests this is now possible, though this is not yet available for routine pathology applications [54]. Other confounding factors not yet known are the variation of the timing of onset of receptivity within one woman between cycles, between women, and/or even whether a woman attains receptivity in every cycle – repeat samples from the same women would provide valuable information but would place a great burden on study participants making collection difficult.

3.5.4 *Time of Sampling*

The first imperative is that sampling for the test does not require additional visits to the clinical specialist. In most instances, attendance is for: (1) early assessment of efficacy of gonadotrophin stimulation, (2) ovum pickup and (3) embryo transfer. Testing could be performed on any of these occasions.

If sampling is performed at early workup, an appointment could be arranged to coincide with the mid-secretory phase of the preceding natural cycle. At this time, testing could establish whether or not endometrial receptivity is achieved during the woman's normal menstrual cycle. If normal, the woman could proceed to IVF. However, it must be taken into account that receptivity may be significantly altered in the stimulated versus the natural cycle [46]. If receptivity is not normally achieved, luteal phase supplementation may be favoured, and the test would enable assessment of its effectiveness.

A test performed following sampling at ovum pickup, the pre-receptive phase, would need to be able to predict development of receptivity and therefore likely pregnancy outcome, if fresh embryo transfer followed. It would enable decision-making as to whether to transfer in that cycle (if that was the woman's preference) or to freeze all embryos and transfer in a 'natural' cycle in which receptivity could be predetermined. The time between sampling at pickup and embryo transfer (usually at 5 days) should enable assay and reporting of the data – such assessments therefore need to be relatively rapid and robust, not requiring complex or lengthy methods and be suitable for use in any IVF/pathology clinic worldwide. However, should current data supporting 'freeze-all' strategies [48] be fully validated in internationally organised trials, this may become the most common protocol in IVF clinics.

It is much less likely that testing at the time of transfer would be useful, unless the test could be performed immediately on site in the clinic. In this situation, a very rapid test yielding a 'yes' or 'no' answer only is required as the time for decision-making and consultation with the patient is limited. Sampling from the uterus, particularly tissue biopsies, would not be appropriate at this time due to possible interference with the implantation environment.

Options for an endometrial receptivity test are summarised in Fig. 3.4.

Fig. 3.4 Options for sampling for an endometrial receptivity test

Assessing receptivity. What is most useful?

- What to sample?
  - Blood, tissue, uterine fluid, urine
- What to measure?
  - mRNA, protein, lipids
  - Single or multiple markers
- When to sample?
  - Mid-secretory in 'normal' cycle
  - LH/hCG+2 in IVF cycle
  - Day of embryo transfer

3.5.5 Transcriptomic Studies

As summarised above, early transcriptomic studies on human endometrium failed to achieve consensus on a molecular signature of receptivity. The studies fall into different categories: (1) those that focussed entirely on normal tissue from women of known fertility, (2) those comparing natural and stimulated cycles [55, 56], (3) studies comparing mid-secretory tissue from fertile versus infertile women [57] and (4) tissue from women with recurrent implantation failure [58, 59]. In terms of biomarker discovery, those studies comparing non-receptive (proliferative or early secretory phase) with receptive (mid-secretory phase) endometrium [60–68] are the most relevant. The differentially expressed genes are both up- and downregulated in the receptive phase. Many relate to known processes of cellular function at this time: cell adhesion, response to external stimuli, signalling, immune response, metabolism and cell-cell communication. Appropriately, there is decreased transcription of genes driving cellular proliferation and development, given that these processes are dominant in the proliferative phase. Details of these findings have been tabulated elsewhere [55, 56].

An important reason for the diversity of results in the studies detailed above is that the endometrial tissue biopsies used for analysis contain a diversity of cell types. In confirmation of this, microarray analysis following laser capture of cellular compartments showed distinct mRNA signatures for glands and stroma, each dependent on the day of the cycle [9]. Unfortunately, this approach is exceptionally time-consuming and not appropriate for routine use as a diagnostic tool for predicting receptivity. It does, however, highlight the redundancy of 'whole tissue' analysis.

An endometrial receptivity array (ERA) [69] is currently undergoing multicentre trials. For this test 238 genes that are differentially regulated within the endometrial cycle are customised on a single array. Of these, 134 genes represent a specific transcriptomic signature of the receptive phase. This test is of high specificity (0.8857) and sensitivity (0.9976) for endometrial dating and is being applied in clinical settings to establish accurately the timing of receptivity in individual women, enabling replacement of a blastocyst at an optimal time. The outcomes of the multicentre trial, particularly in terms of enhancing percentages of live births per cycle, are awaited. This test cannot be performed to predict endometrial receptivity in the same cycle as sampling.

3.5.6 Proteomic Studies

New developments in proteomic technologies provide the ability to detect relatively low-abundance proteins and have enabled considerable advances in the discovery of protein biomarkers. While early studies using gel-based proteomics provided changes in proteins of high abundance, predominantly structural proteins in tissue samples, the newer technologies are enabling identification of much lower-abundance proteins (particularly in biological fluids) that are part of important regulatory pathways. Proteins are the functional mediators of physiological changes, and there are a number of regulated steps between transcription and production of functional protein [55, 70, 71]. These include restriction of translation by miRNAs and post-translation modification of proteins by enzymatic processing, leading to activation or inactivation, glycosylation or phosphorylation. For example, the actions of proprotein convertase 5/6 are essential for receptivity and implantation [72, 73], at least in part by cleavage of a range of proteins including dystroglycan [74], caldesmon [75] and EPB50 [73]. The correlation between abundance of a transcript and its functional protein in the endometrium is often low, and thus examination of transcription by gene array can be misleading in terms of understanding function, although this is not necessarily relevant in provision of biomarkers. Application of proteomic techniques to endometrial biopsies and validation of proteins with relevance to receptivity has been undertaken by a number of research groups [76]: most of these proteins have not been further examined for their potential to assist in a clinical infertility setting.

Uterine fluid, the protein-rich histotroph within the uterine cavity, provides the microenvironment for implantation, including secretions from uterine glands and endometrial luminal epithelium in addition to factors from the Fallopian tubes and blood transudates. Glandular secretions are essential for implantation in both sheep and mice [77, 78]: this has been functionally demonstrated in animals in which uterine gland development was inhibited during early postnatal life. Thus it is predicted that uterine fluid in women will contain important secreted proteins from uterine glands which may be actively involved in the blastocyst implantation process and that could be measured to assess uterine receptivity. This approach has proven very beneficial: proteomic and multiplex analyses of a cohort of factors have identified a number of proteins with considerable potential as markers of receptivity [55, 79]. It is implausible that all women currently defined as having idiopathic infertility have an endometrial-based infertility and, indeed, that those who do will have a single molecular change. Thus it is unlikely that any single biomarker will provide a definitive diagnosis of receptivity or infertility. Future studies will serve to validate and combine these markers using multivariate analysis and will incorporate physical measures (e.g. age, BMI, endometrial thickness) to provide a diagnostic fit of each woman to a spectrum of endometrial defects and produce predictive indices for IVF outcome. Such cohorts of markers will provide utility to monitor response to treatment regimens and provide a more personalised understanding of endometrial receptivity and therapy for infertile women.

The power of multiplex analysis of proteins to distinguish between fertile and infertile women during the mid-secretory phase is demonstrated in Fig. 3.5. Uterine lavage was performed during natural cycles in women <43 years of age who were of proven fertility ($n = 17$) or infertility (<3 failed IVF cycles; $n = 19$). Male factor and tubal and ovarian pathologies were exclusion factors and dating was by Noyes criteria. Ten analytes secreted by endometrial epithelium and previously reported in the literature as potential biomarkers of receptivity were measured. Of the

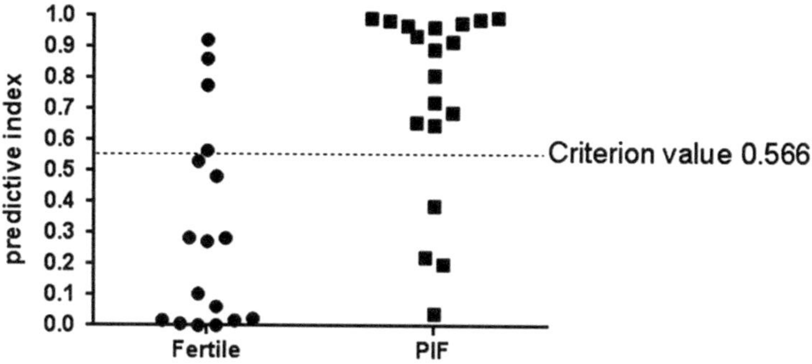

Fig. 3.5 The predictive index of receptivity in the mid-secretory phase of a natural cycle for a cohort of eight protein biomarkers in uterine fluid is shown for women of proven fertility (fertile) and those with primary infertility (PIF) who had >3 unsuccessful IVF cycles. The criterion value is 0.566

individual analytes, only four significantly discriminated between fertile and infertile women. Use of multivariate analysis combining eight of the markers distinguished between fertile and infertile women, with a sensitivity of 79 %, specificity of 82 % and significance of $P < 0.0001$. Of course, the likelihood of a 100 % specificity and sensitivity test is low given the acknowledged potential multiforms of endometrial failure and that not all these women will have idiopathic infertility due to endometrial disturbance but rather some other undiagnosed condition. Indeed, as shown in Fig. 3.5, some primary infertile (PIF) women fall below the cut-off, thus apparently having normal receptive endometrium (based on this set of markers), while a small number of fertile women are predicted as non-receptive being positioned above the threshold. These women's endometrium may be incorrectly pathology dated, given the known inaccuracy of Noyes criteria. This could be determined more accurately with the ERA test discussed above. Further, it is still unproven whether previously fertile women (used as fertile controls) remain fertile in every cycle. Indeed, given the acknowledged relationships of environment, BMI, age and endometriosis with infertility, it is highly plausible that women who have been fertile may subsequently develop infertility. However, this is impossible to determine.

3.5.7 Lipidomics

In mice, double inactivation of cyclooxygenase (COX)-1 and COX-2 completely prevents pregnancy by blocking prostaglandin production and inhibiting implantation [80], providing a rationale for examination of prostaglandins in the uterine cavity of fertile women. PGE_2 and $PGF_2\alpha$ assays have been performed on uterine fluid samples ($n = 39$) at different stages of natural menstrual cycles. In an initial experiment ($n - 13$), PGE_2 and $PGF_2\alpha$ were significantly increased between days

19 and 21 of the cycle – other lipids examined did not change. These data were replicated in a second sample set ($n = 26$). In the combined data set, PGE_2 showed a twofold increase and $PGF_2\alpha$, a 20-fold increase at their peak which coincided with receptivity [81]. A further larger study of endometrial fluid from 173 women [82] added to the evidence that lipid profiling may hold potential. Whether or not these lipids can be used as biomarkers for receptive endometrium remains to be established. Given that prostaglandins are highly unstable, careful sample preparation and storage guidelines are critical if their measurement is to prove clinically and biologically meaningful. Furthermore, the need for laboratory techniques/instrumentation beyond the norm (e.g. gas chromatography and mass spectrometry) may see lipidomics restricted at least in the short term, to a handful of specialist centres.

3.6 Conclusions

While many potential biomarkers for endometrial receptivity have been identified, international collaboration is now needed to adequately validate an optimal cohort of predictive biomarkers and provide a robust test. Standard operating procedures for sampling and storage need to be simplified for use in clinics worldwide. The impact of receptivity testing on clinical outcomes, including assessment of infertility and the impact of a predictive test on decision-making and thus the outcomes of IVF cycles, also remains to be established. Simple, rapid within-clinic testing will be an imperative for a major impact on pregnancy and live birth outcomes. Given the increasing numbers of couples presenting with infertility, such predictive testing to optimise pregnancy success is urgently required. This will reduce costs, both economic and emotional, and provide the best outcomes in terms of take-home healthy babies.

Acknowledgments Work in the authors' laboratory is supported by the National Health and Medical Research Council of Australia by Fellowship (#1002028) and project (#1047056) grants, the Monash IVF Research and Education Foundation, the Merck Serono grants for Fertility Innovation and the Victorian Government's Operational Infrastructure Program.

References

1. Betteridge KJ. An historical look at embryo transfer. J Reprod Fertil. 1981;62(1):1–13.
2. Rowson LE, Moor RM. Embryo transfer in the sheep: the significance of synchronizing oestrus in the donor and recipient animal. J Reprod Fertil. 1966;11(2):207–12.
3. Psychoyos A. Uterine receptivity for nidation. Ann N Y Acad Sci. 1986;476:36–42.
4. Hertig A, Rock J, Adams E. A description of 34 human ova within the first 17 days of development. Am J Anat. 1956;98(3):435–93.

5. Navot D, Bergh PA, Williams M, Garrisi GJ, Guzman I, Sandler B, et al. An insight into early reproductive processes through the in vivo model of ovum donation. J Clin Endocrinol Metab. 1991;72(2):408–14.
6. Wilcox AJ, Baird DD, Weinberg CR. Time of implantation of the conceptus and loss of pregnancy. N Engl J Med. 1999;340(23):1796–9.
7. Tierney EP, Tulac S, Huang ST, Giudice LC. Activation of the protein kinase A pathway in human endometrial stromal cells reveals sequential categorical gene regulation. Physiol Genomics. 2003;16(1):47–66.
8. Murray MJ, Meyer WR, Zaino RJ, Lessey BA, Novotny DB, Ireland K, et al. A critical analysis of the accuracy, reproducibility, and clinical utility of histologic endometrial dating in fertile women. Fertil Steril. 2004;81(5):1333–43.
9. Evans GE, Martinez-Conejero JA, Phillipson GT, Simon C, McNoe LA, Sykes PH, et al. Gene and protein expression signature of endometrial glandular and stromal compartments during the window of implantation. Fertil Steril. 2012;97(6):1365–73.
10. Salamonsen LA, Nie G, Hannan NJ, Dimitriadis E, Society for Reproductive Biology Founders' Lecture. Preparing fertile soil: the importance of endometrial receptivity. Reprod Fertil Dev. 2009;21(7):923–34.
11. Gellersen B, Brosens J. Cyclic AMP and progesterone receptor cross-talk in human endometrium: a decidualizing affair. J Endocrinol. 2003;178(3):357–72.
12. Brosens J, Salker M, Teklenburg G, Nautiyal J, Salter S, Lucas E, et al. Uterine selection of human embryos at implantation. Sci Rep. 2014;4:3984.
13. Evans J, Salamonsen L, Menkhorst E, Dimitriadis E. Dynamic changes in hyperglycosylated human chorionic gonadotrophin throughout the first trimester of pregnancy and its role in early placentation. Hum Reprod. 2015;30(5):1029–38.
14. Weimar C, Kavelaars A, Brosens J, Gellersen B, de Vreeden-Elbertse J, Heijnen C, et al. Endometrial stromal cells of women with recurrent miscarriage fail to discriminate between high- and low-quality human embryos. PLoS One. 2012;7(7):e41424.
15. Teklenburg G, Salker MS, Molokhia M, Lavery S, Trew G, Aojanepong T, et al. Natural selection of human embryos: decidualizing endometrial stromal cells serve as sensors of embryo quality upon implantation. PLoS One. 2010;5(4):e10258.
16. Schwenke M, Knofler M, Velicky P, Weimar C, Kruse M, Samalecos A, et al. Control of human endometrial stromal cell motility by PDGF-BB, HB-EGF and trophoblast-secreted factors. PLoS One. 2013;8(1):e54336.
17. Bonduelle ML, Dodd R, Liebaers I, Van Steirteghem A, Williamson R, Akhurst R. Chorionic gonadotrophin-beta mRNA, a trophoblast marker, is expressed in human 8-cell embryos derived from tripronucleate zygotes. Hum Reprod. 1988;3(7):909–14.
18. Kane N, Kelly RW, Saunders P, Critchley H. Proliferation of uterine natural killer cells is induced by human chorionic gonadotropin and mediated via the mannose receptor. Endocrinology. 2009;150(6):2882–8.
19. Licht P, Fluhr H, Neuwinger J, Wallwiener D, Wildt L. Is human chorionic gonadotropin directly involved in the regulation of human implantation? Mol Cell Endocrinol. 2007;269(1–2):85–92.
20. Evans J, Salamonsen L. Too much of a good thing? Experimental evidence suggests prolonged exposure to hCG is detrimental to endometrial receptivity. Hum Reprod. 2013;28(6):1610–9.
21. Lopata A, Oliva K, Stanton PG, Robertson DM. Analysis of chorionic gonadotrophin secreted by cultured human blastocysts. Mol Hum Reprod. 1997;3(6):517–21.
22. Licht P, Losch A, Dittrich R, Neuwinger J, Siebzehnrubl E, Wildt L. Novel insights into human endometrial paracrinology and embryo-maternal communication by intrauterine microdialysis. Hum Reprod Update. 1998;4(5):532–8.
23. Sherwin JR, Sharkey AM, Cameo P, Mavrogianis PM, Catalano RD, Edassery S, et al. Identification of novel genes regulated by chorionic gonadotropin in baboon endometrium during the window of implantation. Endocrinology. 2007;148(2):618–26.

24. Paiva P, Hannan NJ, Hincks C, Meehan KL, Pruysers E, Dimitriadis E, et al. Human chorionic gonadotrophin regulates FGF2 and other cytokines produced by human endometrial epithelial cells, providing a mechanism for enhancing endometrial receptivity. Hum Reprod. 2011;26 (5):1153–62.
25. Evans J, Catalano RD, Brown P, Sherwin R, Critchley HO, Fazleabas AT, et al. Prokineticin 1 mediates fetal-maternal dialogue regulating endometrial leukemia inhibitory factor. FASEB J. 2009;23(7):2165–75.
26. Thouas G, Dominguez F, Green M, Vilella F, Simon C, Gardner D. Soluble ligands and their receptors in human embryo development and implantation. Endocr Rev. 2015;36(1):92–130.
27. Robertson S, Chin P, Schjenken J, Thompson J. Female tract cytokines and developmental programming in embryos. In: Leese H, Brison D, editors. Cell signalling during mammalian early embryo development, Advances in Experimental Medicine and Biology, vol. 843. New York: Springer; 2015. p. 173–213.
28. Ziebe S, Loft A, Povlsen B, Erb K, Agerholm I, Aasted M, et al. A randomized clinical trial to evaluate the effect of granulocyte-macrophage colony-stimulating factor (GM-CSF) in embryo culture medium for in vitro fertilization. Fertil Steril. 2013;99:1600–9.
29. Hannan NJ, Paiva P, Meehan KL, Rombauts LJ, Gardner DK, Salamonsen LA. Analysis of fertility-related soluble mediators in human uterine fluid identifies VEGF as a key regulator of embryo implantation. Endocrinology. 2011;152(12):4948–56.
30. Binder N, Evans J, Gardner D, Salamonsen L, Hannan N. Endometrial signals improve embryo outcome: functional role of vascular endothelial growth factor isoforms on embryo development and implantation in mice. Hum Reprod. 2014;29(10):2278–86.
31. Macaldowie A, Wang Y, Chambers G, Sullivan E. Assisted reproductive technology in Australia and New Zealand 2011. Sydney: National Perinatal Epidemiology and Statistics Unit, The University of New South Wales, Australia; 2013.
32. Ferraretti A, Goossens V, Kupka M, Bhattacharya S, de Mouzon J, Castilla J, et al. Assisted reproductive technology in Europe, 2009: results generated from European registers by ESHRE. Hum Reprod. 2013;28(9):2318–31.
33. Society for Assisted Reproductive Technology: IVF success rate reports. http://www.sart.org. 2011.
34. Venetis C, Kolibianakis E, Bosdou J, Tarlatzis B. Progesterone elevation and probability of pregnancy after IVF: a systematic review and meta-analysis of over 60,000 cycles. Hum Reprod Update. 2013;19:433–57.
35. Labarta E, Martinez-Conejero J, Alama P, Horcajadas J, Pellicer A, Simon C, et al. Endometrial receptivity is affected in women with high circulating progesterone levels at the end of the follicular phase: a functional genomics analysis. Hum Reprod. 2011;26 (7):1813–25.
36. Li R, Qiao J, Wang L, Li L, Zhen X, Liu P, et al. MicroRNA array and microarray evaluation of endometrial receptivity in patients with high serum progesterone levels on the day of hCG administration. Reprod Biol Endocrinol. 2011;9:29.
37. Haouzi D, Bissonnette L, Gala A, Assou S, Entezami F, Perrochia H, et al. Endometrial receptivity profile in patients with premature progesterone elevation on the day of HCG administration. Biomed Res Int. 2014;epub Apr 28.
38. Legro R, Ary B, Paulson R, Stanczyk F, Sauer M. Premature lutenization as detected by elevated serum progesterone is associated with a higher pregnancy rate in donor oocyte in-vitro fertilization. Hum Reprod. 1993;8(9):1506–11.
39. Melo M, Meseguer M, Garrido N, Bosch E, Pellicer A, Remohi J. The significance of premature lutenization in an oocyte-donation programme. Hum Reprod. 2006;21(6):1503–7.
40. Noyes RW, Hertig AT, Rock J. Dating the endometrial biopsy. Am J Obstet Gynecol. 1975;122(2):262–3.
41. Chetkowski R, Kiltz R, Salyer W. In premature luteinization, progesterone induces secretory transformation of the endometrium without impairment of embryo viability. Fertil Steril. 1997;68(2):292–7.

42. Ubaldi F, Bourgain C, Tournaye H, Smitz J, Van Steirteghem A, Devroey P. Endometrial evaluation by aspiration biopsy on the day of oocyte retrieval in the embryo transfer cycles in patients with serum progesterone rise during the follicular phase. Fertil Steril. 1997;67 (3):521–6.
43. Kolibianakis EM, Devroey P. The luteal phase after ovarian stimulation. Reprod Biomed Online. 2002;5(Suppl 1(3)):26–35.
44. Van Vaerenbergh I, Van Lommel L, Ghislain V, In't Veld P, Schuit F, Fatemi HM, et al. In GnRH antagonist/rec-FSH stimulated cycles, advanced endometrial maturation on the day of oocyte retrieval correlates with altered gene expression. Hum Reprod. 2009;24(5):1085–91.
45. Horcajadas J, Minguez P, Dopazo J, Esteban F, Dominquez F, Giudice L, et al. Controlled ovarian stimulation induces a functional genomic delay of the endometrium with potential clinical implications. J Clin Endocrinol Metab. 2008;93(11):4500–10.
46. Evans J, Hannan NJ, Hincks C, Rombauts LJ, Salamonsen LA. Defective soil for a fertile seed? Altered endometrial development is detrimental to pregnancy success. PLoS One. 2012;7(12):e53098.
47. Shapiro B, Daneshmand S, Garner F, Aguirre M, Hudson C. Freeze-all can be a superior therapy to another fresh cycle in patients with prior fresh blastocyst implantation failure. Reprod Biomed Online. 2014;29(3):286–90.
48. Evans J, Hannan N, Edgell T, Vollenhoven B, Lutjen P, Osianlis T, et al. Fresh versus frozen embryo transfer: backing clinical decisions with scientific and clinical evidence. Hum Reprod Update. 2014;20(6):808–21.
49. Altmae S, Esteban F, Stavreus-Evers A, Simon C, Giudice LC, Lessey B, et al. Guidelines for the design, analysis and interpretation of 'omics' data: focus on human endometrium. Hum Reprod Update. 2014;20(1):12–28.
50. Nastri C, Lensen S, Gibreel A, Raine-Fenning N, Ferriani R, Bhattacharya S, et al. Endometrial injury in women undergoing assisted reproductive techniques. Cochrane Database Syst Rev. 2015;3:CD009517.
51. Hannan NJ, Nie G, Rainzcuk A, Rombauts LJ, Salamonsen LA. Uterine lavage or aspirate: which view of the intrauterine environment? Reprod Sci. 2012;19:1125–32.
52. van der Gaast MH, Beier-Hellwig K, Fauser BC, Beier HM, Macklon NS. Endometrial secretion aspiration prior to embryo transfer does not reduce implantation rates. Reprod Biomed Online. 2003;7(1):105–9.
53. Diaz-Gimeno P, Ruiz-Alonso M, Blesa D, Bosch N, Martinez-Conejero JA, Alama P, et al. The accuracy and reproducibility of the endometrial receptivity array is superior to histology as a diagnostic method for endometrial receptivity. Fertil Steril. 2013;99(2):508–17.
54. Lessey BA. The pathologists are free to go, or are they? Fertil Steril. 2013;99(2):350–1.
55. Haouzi D, Dechaud H, Assou S, De Vos J, Hamamah S. Insights into human endometrial receptivity from transcriptomic and proteomic data. Reprod Biomed Online. 2012;24 (1):23–34.
56. Ruiz-Alonso M, Blesa D, Simon C. The genomics of the human endometrium. Biochim Biophys Acta. 2012;1822(12):1931–42.
57. Altmae S, Martinez-Conejero JA, Salumets A, Simon C, Horcajadas JA, Stavreus-Evers A. Endometrial gene expression analysis at the time of embryo implantation in women with unexplained infertility. Mol Hum Reprod. 2010;16(3):178–87.
58. Ledee N, Munaut C, Aubert J, Serazin V, Rahmati M, Chaouat G, et al. Specific and extensive endometrial deregulation is present before conception in IVF/ICSI repeated implantation failures (IF) or recurrent miscarriages. J Pathol. 2011;225(4):554–64.
59. Othman R, Omar MH, Shan LP, Shafiee MN, Jamal R, Mokhtar NM. Microarray profiling of secretory-phase endometrium from patients with recurrent miscarriage. Reprod Biol. 2012;12 (2):183–99.
60. Borthwick JM, Charnock-Jones DS, Tom BD, Hull ML, Teirney R, Phillips SC, et al. Determination of the transcript profile of human endometrium. Mol Hum Reprod. 2003;9(1):19–33.

61. Carson DD, Lagow E, Thathiah A, Al-Shami R, Farach-Carson MC, Vernon M, et al. Changes in gene expression during the early to mid-luteal (receptive phase) transition in human endometrium detected by high-density microarray screening. Mol Hum Reprod. 2002;8 (9):871–9.
62. Haouzi D, Assou S, Mahmoud K, Tondeur S, Reme T, Hedon B, et al. Gene expression profile of human endometrial receptivity: comparison between natural and stimulated cycles for the same patients. Hum Reprod. 2009;24(6):1436–45.
63. Kao LC, Tulac S, Lobo S, Imani B, Yang JP, Germeyer A, et al. Global gene profiling in human endometrium during the window of implantation. Endocrinology. 2002;143 (6):2119–38.
64. Kuokkanen S, Chen B, Ojalvo L, Benard L, Santoro N, Pollard JW. Genomic profiling of microRNAs and messenger RNAs reveals hormonal regulation in microRNA expression in human endometrium. Biol Reprod. 2010;82(4):791–801.
65. Ponnampalam AP, Weston GC, Trajstman AC, Susil B, Rogers PA. Molecular classification of human endometrial cycle stages by transcriptional profiling. Mol Hum Reprod. 2004;10 (12):879–93.
66. Riesewijk A, Martin J, van Os R, Horcajadas JA, Polman J, Pellicer A, et al. Gene expression profiling of human endometrial receptivity on days LH+2 versus LH+7 by microarray technology. Mol Hum Reprod. 2003;9(5):253–64.
67. Talbi S, Hamilton AE, Vo KC, Tulac S, Overgaard MT, Dosiou C, et al. Molecular phenotyping of human endometrium distinguishes menstrual cycle phases and underlying biological processes in normo-ovulatory women. Endocrinology. 2006;147(3):1097–121.
68. Tseng LH, Chen I, Chen MY, Yan H, Wang CN, Lee CL. Genome-based expression profiling as a single standardized microarray platform for the diagnosis of endometrial disorder: an array of 126-gene model. Fertil Steril. 2010;94(1):114–9.
69. Diaz-Gimeno P, Horcajadas JA, Martinez-Conejero JA, Esteban FJ, Alama P, Pellicer A, et al. A genomic diagnostic tool for human endometrial receptivity based on the transcriptomic signature. Fertil Steril. 2011;95(1):50–60.
70. Burney RO, Talbi S, Hamilton AE, Vo KC, Nyegaard M, Nezhat CR, et al. Gene expression analysis of endometrium reveals progesterone resistance and candidate susceptibility genes in women with endometriosis. Endocrinology. 2007;148(8):3814–26.
71. Chen JI, Hannan NJ, Mak Y, Nicholls PK, Zhang J, Rainczuk A, et al. Proteomic characterization of midproliferative and midsecretory human endometrium. J Proteome Res. 2009;8 (4):2032–44.
72. Nie G, Li Y, Wang M, Liu YX, Findlay JK, Salamonsen LA. Inhibiting uterine PC6 blocks embryo implantation: an obligatory role for a proprotein convertase in fertility. Biol Reprod. 2005;72(4):1029–36.
73. Heng S, Cervero A, Simon C, Stephens AN, Li Y, Zhang J, et al. Proprotein convertase 5/6 is critical for embryo implantation in women: regulating receptivity by cleaving EBP50, modulating ezrin binding, and membrane-cytoskeletal interactions. Endocrinology. 2011;152 (12):5041–52.
74. Heng S, Paule S, Ying L, Rombauts L, Vollenhoven B, Salamonsen L, et al. Post-translational removal of α-dystroglycan N-terminus by PC5/6 cleavage is important for uterine preparation for embryo implantation in women. FASEB J. 2015;29(9):4011–22.
75. Kilpatrick LM, Stephens AN, Hardman BM, Salamonsen LA, Li Y, Stanton PG, et al. Proteomic identification of caldesmon as a physiological substrate of proprotein convertase 6 in human uterine decidual cells essential for pregnancy establishment. J Proteome Res. 2009;8(11):4983–92.
76. Edgell T, Rombauts L, Salamonsen L. Assessing receptivity in the endometrium: the need for a rapid, non-invasive test. Reprod Biomed Online. 2013;27(5):486–96.
77. Gray CA, Bartol FF, Tarleton BJ, Wiley AA, Johnson GA, Bazer FW, et al. Developmental biology of uterine glands. Biol Reprod. 2001;65(5):1311–23.

78. Dunlap KA, Filant J, Hayashi K, Rucker 3rd EB, Song G, Deng JM, et al. Postnatal deletion of Wnt7a inhibits uterine gland morphogenesis and compromises adult fertility in mice. Biol Reprod. 2011;85(2):386–96.
79. Hannan NJ, Stephens AN, Rainczuk A, Hincks C, Rombauts LJ, Salamonsen LA. 2D-DiGE analysis of the human endometrial secretome reveals differences between receptive and nonreceptive states in fertile and infertile women. J Proteome Res. 2010;9(12):6256–64.
80. Reese J, Zhao X, Ma W, Brown N, Maziasz T, Dey S. Comparative analysis of pharmacologic and/or genetic disruption of cyclooxygenase-1 and cyclooxygenase-2 function in female reproduction in mice. Endocrinology. 2001;142(7):3198–206.
81. Berlanga O, Bradshaw H, Vilella-Mitjana F, Garrido-Gomez T, Simon C. How endometrial secretomics can help in predicting implantation. Placenta. 2011;32 Suppl 3:S271–5.
82. Vilella F, Ramirez L, Berlanga O, Martinez S, Alama P, Mesequer M, et al. PGE2 and PGF2*a* concentrations in human endometrial fluid as biomarkers for embryonic implantation. J Clin Endocrinol Metab. 2013;98(10):4123–32.
83. Heng S, Hannan NJ, Rombauts LJ, Salamonsen LA, Nie G. PC6 levels in uterine lavage are closely associated with uterine receptivity and significantly lower in a subgroup of women with unexplained infertility. Hum Reprod. 2011;26(4):840–6.
84. Lessey BA, Castelbaum AJ, Sawin SW, Sun J. Integrins as markers of uterine receptivity in women with primary unexplained infertility. Fertil Steril. 1995;63(3):535–42.
85. Dominguez F, Garrido-Gomez T, Lopez JA, Camafeita E, Quinonero A, Pellicer A, et al. Proteomic analysis of the human receptive versus non-receptive endometrium using differential in-gel electrophoresis and MALDI-MS unveils stathmin 1 and annexin A2 as differentially regulated. Hum Reprod. 2009;24(10):2607–17.
86. Garrido-Gomez T, Quinonero A, Antunez O, Diaz-Gimeno P, Bellver J, Simon C, et al. Deciphering the proteomic signature of human endometrial receptivity. Hum Reprod. 2014;29(9):1957–67.
87. van der Gaast MH, Macklon NS, Beier-Hellwig K, Krusche CA, Fauser BC, Beier HM, et al. The feasibility of a less invasive method to assess endometrial maturation – comparison of simultaneously obtained uterine secretion and tissue biopsy. Br J Obstet Gynecol. 2009;116 (2):304–12.
88. Scotchie J, Fritz M, Mocanu M, Lessey B, Young SL. Proteomic analysis of the luteal endometrial secretome. Reprod Sci. 2009;16(9):883–93.

Chapter 4
Role of Circulating Blood Cells in Maternal Tissue Remodeling and Embryo-Maternal Cross Talk

Hiroshi Fujiwara, Yoshihiko Araki, Yukiyasu Sato, Masahiro Takakura, Mitsuhiro Nakamura, Yasunari Mizumoto, Akihito Horie, Hideharu Kanzaki, and Takahide Mori

Abstract In humans, progesterone production of the corpus luteum during pregnancy is maintained by human chorionic gonadotropin (HCG) secreted by the implanting embryo. In addition to this endocrine system, accumulating evidence suggests that circulating immune cells play an important role in the embryo-maternal cross talk through blood circulation. Peripheral blood mononuclear cells (PBMC) derived from women in early pregnancy enhanced progesterone production by human luteal cells and trophoblast invasion in vitro. Spleen cells derived from mice during early pregnancy induced endometrial differentiation and embryo implantation in vivo. Furthermore, recombinant HCG stimulated human PBMC to produce chemokines through lectin-glycan interaction, promoting trophoblast invasion. Consequently, we proposed that the maternal immune system undergoes functional changes by recognizing developing embryos from the early stage of pregnancy and assists embryo implantation in cooperation with the endocrine system. On the other hand, chemokines secreted from the locally deposited platelets

---

H. Fujiwara (✉) • M. Takakura • M. Nakamura • Y. Mizumoto
Department of Obstetrics and Gynecology, Kanazawa University Graduate School of Medical Science, 13-1 Takaramachi, Kanazawa, Ishikawa, Japan
e-mail: fuji@kuhp.kyoto-u.ac.jp

Y. Araki
Institute for Environmental & Gender-specific Medicine, Juntendo University Graduate School of Medicine, 2-1-1 Tomioka, Urayasu-City 279-0021, Japan

Y. Sato
Department of Obstetrics and Gynecology, Otsu Red Cross Hospital, Shiga, Japan

A. Horie
Department of Gynecology and Obstetrics, Kyoto University Graduate School of Medicine, Kyoto, Japan

H. Kanzaki
Department of Obstetrics and Gynecology, Kansai Medical University, Osaka, Japan

T. Mori
Academia for Repro-Regenerative Medicine, Tokyo, Japan

H. Kanzaki (ed.), *Uterine Endometrial Function*,
DOI 10.1007/978-4-431-55972-6_4

induced neovascularization during corpus luteum formation and promoted extravillous trophoblast invasion during placental formation, reconstructing maternal endometrial spiral arteries. These findings suggest that circulating blood cells including PBMC and platelets positively contribute to embryo-maternal cross talk and maternal tissue remodeling around the implantation period.

Keywords Corpus luteum • Embryo implantation • Endometrial differentiation • HCG • PBMC

4.1 Introduction

The uterus is a unique organ that can receive embryo implantation. Consequently, the mammalian mother must interact with the implanting embryo in the female genital tract and reconstruct maternal tissues during placental formation, adapting maternal organs to accept the embryo in the uterus. To induce functional changes of the uterine environment to receive embryo implantation, mammalian females utilize progesterone, constructing a progesterone-producing organ, the corpus luteum (CL), from the ovulated follicle in the ovary. Progesterone induces the endometrium to become suitable for embryo implantation. It is also widely accepted that direct cross talk between the embryo and maternal endometrium during the migration of the developing embryo through the oviduct into the uterine cavity is necessary to achieve successful implantation [1]. However, precise mechanisms of human embryo-maternal cross talk and subsequent embryo implantation remain unknown. Current evidence suggests that local immune cells at the implantation site actively contribute to embryo implantation [2–4].

Recently, we demonstrated evidence that circulating blood immune cells contribute to embryo implantation by regulating the CL function and endometrial differentiation and that platelets are involved in the formation of the CL and placenta by remodeling maternal tissues. In this chapter, we describe our new concepts concerning the physiological roles of circulating blood cells in embryo-maternal cross talk and maternal tissue remodeling.

4.2 Regulation of Female Reproductive Organs by Circulating Blood Cells

4.2.1 Regulation of CL Formation by PBMC and Platelets

It has been proposed that ovulation is an inflammatory reaction caused by various proteolytic enzymes such as serine proteases and metalloproteases [5]. During ovulation, granulosa cells undergo luteinization and transform into large luteal cells that produce abundant progesterone, while theca interna cells become small luteal cells [6]. In parallel with the luteinizing process, endothelial cells in the theca interna layer initiate migration into the granulosa cell layer, constructing a new vascular network among luteinizing granulosa cells and finally achieving vascular anastomosis in the central cavity area [7]. Several soluble angiogenic factors such as vascular endothelial growth factor (VEGF), angiogenin, endocrine gland-VEGF, and angiopoietin were demonstrated to be secreted by luteinizing granulosa cells [8–11]. Both luteinization and neovascularization processes are necessary to establish a mature CL, which is suitably synchronized with endometrial differentiation to receive the well-timed implantation of the developing embryo. When pregnant, the CL of the menstrual cycle is further transformed into the CL of pregnancy. The function of the CL of pregnancy is maintained for more than 2 months, while the functional life span of the CL of the menstrual cycle ceases 14 days after ovulation when embryo implantation does not occur [12].

The differentiation of granulosa cells is considered to be mainly regulated by gonadotropins and also modulated by growth factors and cytokines. Luteinizing hormone (LH) is a key factor regulating the process of CL formation. LH/HCG receptors are expressed on the cell surface of luteinizing granulosa and theca interna cells. We previously reported that luteinizing granulosa cells increase the expression of immunoreactive LH/HCG receptor during CL formation, and that LH/HCG receptor expression on luteal cells is also maintained in the CL of pregnancy, while LH/HCG receptor expression disappears on luteal cells in the regressing CL [13]. HCG promotes progesterone production by luteinizing granulosa cells in vitro. HCG was also reported to increase VEGF production by granulosa cells [14], supporting that LH is one of the major factors that directly induce luteinization and neovascularization in the CL. However, since CL formation is a marked tissue-remodeling event, it is difficult to explain this temporal, spatial, and three-dimensional construction of a vascular-rich endocrine organ only through the endocrine system using pituitary LH alone. Consequently, the precise regulatory mechanisms of human CL formation have not yet been elucidated.

Previously, we demonstrated that certain soluble factors secreted by PBMC, especially T lymphocytes, enhanced progesterone production by cultured human luteinizing granulosa cells [15]. We then reported that cytokines could regulate the luteinization of cultured human granulosa cells [16]. These findings suggest that the circulating immune cells physiologically contribute to the luteinization process of human granulosa cells during CL formation (Fig. 4.1).

On the other hand, as new mechanisms for neovascularization during CL formation, we demonstrated that luteinizing granulosa cells increased the expression of ephrin B1 and melanoma cell adhesion molecule/CD146 on the cell surface, which can regulate endothelial migration and vessel formation via cell-to-cell contact [17–19]. Although the concept that these cell-to-cell interaction-mediating molecules are involved in the vascular network formation within the CL is attractive, these factors cannot clearly explain the centripetal induction of endothelial migration toward the central cavity where the anastomosis is finally achieved.

Later, we found that circulating platelets were deposited and activated in the extravascular spaces among luteinizing granulosa cells during human CL formation

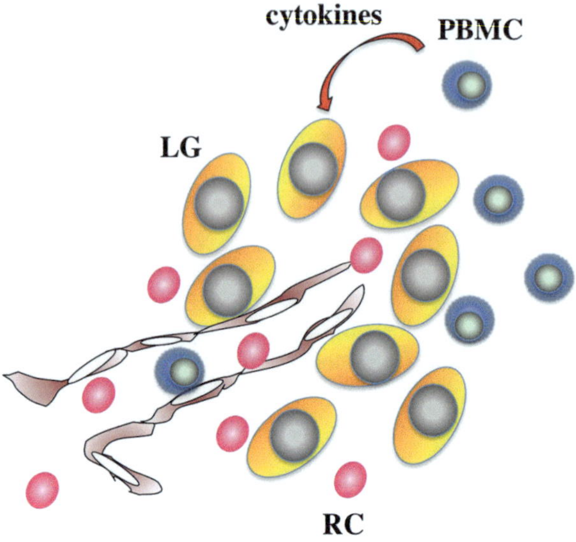

Fig. 4.1 The circulating immune cells promote the luteinization process of human granulosa cells by secreting cytokines during CL formation. *LG* luteinizing granulosa cells, *RC* red cells

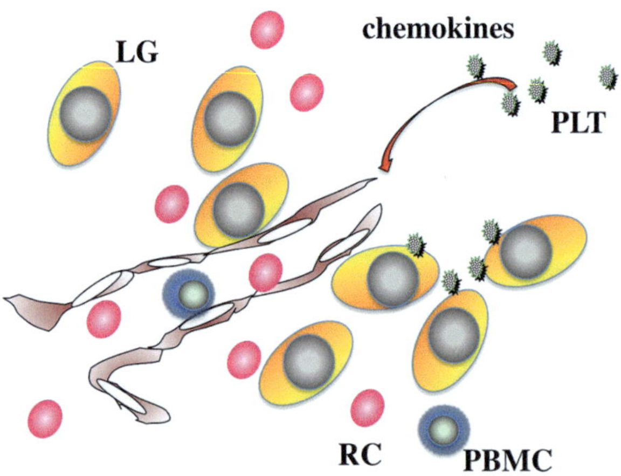

Fig. 4.2 The circulating platelets are deposited and activated in the extravascular spaces among luteinizing granulosa cells during human CL formation. The soluble factors secreted from the activated platelets promote the progesterone production of luteinizing granulosa cells and the migration of endothelial cells. *LG* luteinizing granulosa cells, *PLT* platelets, *RC* red cells

[20]. Platelet deposition gradually became limited near the central cavity toward which microvessels were extending, being established with the vascular network in the mid-luteal phase. The soluble factors secreted from the activated platelets promoted the progesterone production of luteinizing granulosa cells and the migration of human umbilical vein endothelial cells, whereas luteinizing granulosa cells attenuated platelet-induced endothelial cell migration [20], leading to the proposal of the novel concept that platelets are physiological regulators of the remodeling process of the human corpus luteum (Fig. 4.2).

4.2.2 *Regulation of Transformation to CL of Pregnancy by PBMC*

After successful implantation, HCG secreted from the embryo induces transformation from the menstrual cycle into the CL of pregnancy, leading to the maintenance of embryo implantation. Although it has been accepted that HCG is a major regulator of the human CL of pregnancy, there are many lines of clinical evidence suggesting the presence of different regulatory factors for the human CL of pregnancy [21]. However, no soluble factor other than HCG has been identified and the precise regulatory mechanisms remain unknown [22]. To identify new mechanisms, we raised monoclonal antibodies that reacted with luteal cells in the human CL to detect the key molecules expressed on the human CL of pregnancy. Through this project, several molecules such as HLA-DR, leukocyte functional antigen (LFA)-3/CD58, and activated leukocyte adhesion molecules (ALCAM)/CD166, which mediate interaction with T lymphocytes, were found to be expressed on the human luteal cell surface from the stage of CL formation to the CL of pregnancy [23, 24]. Accordingly, in contrast to the previous concept that immune cells enhance CL regression [25], we proposed that immune cells are involved in the transition from the CL of the menstrual cycle to the CL of pregnancy and its functional regulation. In other words, we speculated that signals from the developing embryo in the genital tract are transmitted to the ovary by not only the endocrine system but also the immune system, i.e., via not only soluble factors but also circulating cells [25] (Fig. 4.3).

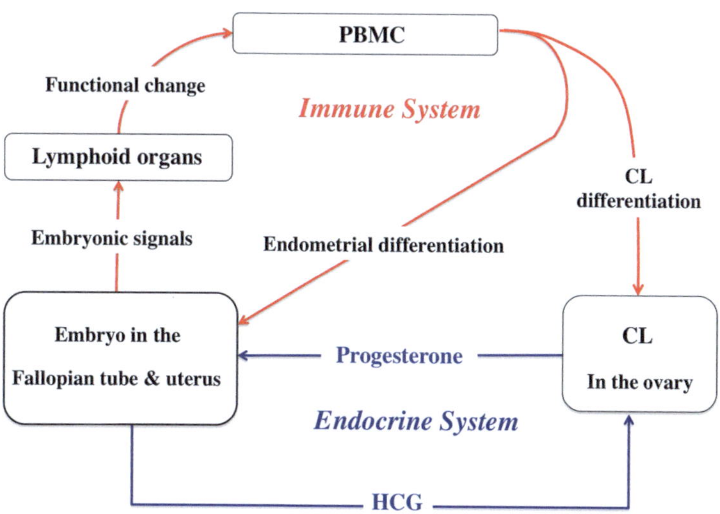

Fig. 4.3 The circulating immune cells change their function on receiving embryonic signals and then induce CL function, early endometrial differentiation, and subsequent embryo implantation, suggesting the presence of dual control through the endocrine and immune systems

To investigate this issue, we examined the effects of PBMC on the luteal cell function in vitro and found that PBMC derived from women in early pregnancy promoted progesterone production by cultured luteal cells isolated from the human CL, suggesting that circulating blood immune cells in women in early pregnancy promote CL function [26]. Furthermore, in the same coculture system, the production of Th-2 cytokines such as IL-4 and IL-10 was enhanced, and these cytokines significantly promoted progesterone production by human luteal cells to the same level as HCG stimulation [26]. These findings support our hypothesis and lead to a further extended novel concept that circulating immune cells transmit information about the presence of the developing embryo to various organs throughout the whole body and induce adequate functional change or differentiation in these organs to facilitate embryo implantation [27].

4.2.3 Regulation of Endometrial Differentiation by Circulating Immune Cells

To verify the above concept concerning the facilitating functions of the circulating immune cells on embryo implantation, we examined the effects of murine spleen cells, stocked circulating immune cells, on endometrial differentiation and embryo implantation using mouse implantation experiments. When blastocysts were transferred into the uterine cavity of pseudopregnant recipient mice that had been mated with vasectomized male mice, successful implantation was observed during 3–5 days after ovulation when the endometrium was adequately differentiated. This period is called the "implantation window" [28, 29]. However, when spleen cells obtained from mice on day 4 of pregnancy were administered to pseudopregnant mice, embryo implantation was induced on 1–2 days after ovulation, when embryos cannot normally be implanted [30]. Furthermore, using a delayed implantation model in which pseudopregnant mice were treated with daily progesterone supplementation following an oophorectomy on postovulatory day 3, we demonstrated that the intravenous administration of splenocytes derived from early pregnancy alone can induce the expression of leukemia inhibitory factor (LIF) in the endometrium and embryo implantation [31], as demonstrated in a previous report that estrogen can induce LIF expression and restart the implantation of an embryo that remains floating in the luminal spaces [32]. In addition, thymocytes from nonpregnant immature mice, especially a CD8-negative population, were demonstrated to induce LIF expression and promote embryo implantation in a delayed implantation model, indicating that an implantation-inducing immune cell population is present even in nonpregnant mice [33]. From these findings, we concluded that circulating murine immune cells induce early endometrial differentiation and subsequent embryo implantation, suggesting the dual control of endometrial differentiation through the endocrine and immune systems (Fig. 4.3).

To examine whether the effects of PBMC on murine endometrial differentiation can correspond to human endometrial receptivity, we then developed an in vitro attachment assay using a primary culture of human endometrial epithelial cells and a human choriocarcinoma-derived BeWo cell mass, where high attachment rates were observed in endometrial epithelial cells derived from the mid-luteal phase. In this assay, when autologous PBMC were cocultured with the endometrial cells derived from women in the late proliferative and early secretory phases, attachment rates were significantly increased, indicating that autologous PBMC can promote endometrial cell receptivity [34].

In recent years, repeated implantation failure in infertile patients receiving in vitro fertilization (IVF) therapy has become one of the most important clinical problems. On the basis of our findings, we developed a novel therapy using autologous PBMC. In this therapy, autologous PBMC are administered into the uterine cavity before blastocyst transfer, expecting adequate endometrial differentiation for embryo implantation. When we applied the treatment using autologous PBMC to patients following repeated implantation failure in IVF therapy, this therapy effectively improved both pregnancy and implantation rates [35], supporting that the dual control of endometrial differentiation through the endocrine and immune systems also operates in human pregnancy.

4.2.4 Direct Effects of HCG on Immune Cells Through Sugar Chain Receptors

The above findings suggest that the immune system recognizes the presence of the embryo in the female genital tract. As a specific signal from the embryo to the immune system, we paid attention to HCG, which is secreted from the developing and implanting human embryo. We first examined the effects of PBMC derived from women in early pregnancy on the invasion of BeWo cells or murine embryos using a matrigel invasion assay. In this assay, PBMC derived from women in early pregnancy promoted both murine trophectoderm and BeWo cell invasion more than PBMC obtained from nonpregnant women. Importantly, when the cultured PBMC derived from nonpregnant women were preincubated with HCG, the promoting effects of PBMC were enhanced by HCG treatment, suggesting that HCG can modify PBMC functions to facilitate embryo implantation [36, 37].

Initially, it was reported that HCG crudely purified from the urine of pregnant women suppressed immune reactions [38]. Later, highly purified HCG was shown to have no effect on lymphocyte function [39]. Accordingly, the effects of HCG on the immune cell function remained unclear for a long time. With this background, we demonstrated that human recombinant HCG produced by murine cell lines enhanced IL-8 production by human monocytes. This promoting effect of HCG was observed at relatively high concentrations and its intracytoplasmic signal was mediated via the activation of NF-kB [40]. Although HCG accesses LH/HCG

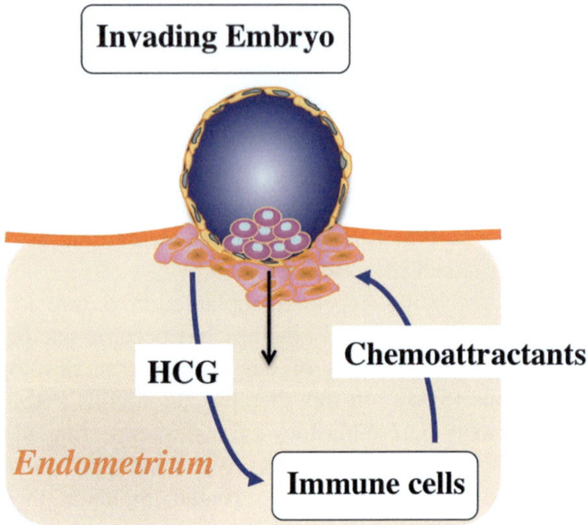

Fig. 4.4 HCG at a high concentration secreted at the embryo implantation site stimulates endometrial immune cells to produce chemoattractants, and then these cytokines, in turn, induce embryo invasion toward the endometrial stroma

receptors, the cell surface expression of LH/HCG receptors was hardly detected on monocytes, suggesting that a different pathway that can respond to high HCG concentrations is present on monocytes. In contrast to LH, chorionic gonadotropins are detected in very limited species, including primates and horses. Consequently, HCG is an evolutionarily new hormone [41], and the most important difference between LH and HCG is the presence of abundant sugar chains at the C-terminal of the HCG β-subunit. In general, the sugar chains of HCG in the blood circulation are largely cleaved in the liver before urine production [42]. When an excess of sugars were added to the culture of HCG-treated monocytes, HCG-induced IL-8 production was inhibited in a dose-dependent manner. These findings suggest that HCG can regulate PBMC function through sugar chain receptors, which is a primitive mechanism of the immune regulatory system [40]. In accordance with our findings, a recent study reported that human invading trophoblasts at the implantation site produces hyperglycosylated HCG and that the hyperglycosylated HCG upregulates trophoblast invasion in humans [43]. Taken together, we speculate that HCG at high concentration secreted at the embryo implantation site stimulates endometrial immune cells to produce chemoattractants and that these cytokines, in turn, induce embryo invasion toward the endometrial stroma [27] (Fig. 4.4).

4.2.5 Regulation of Remodeling of Maternal Endometrial Spinal Artery by Circulating Platelets

After the human embryo migrates within the endometrial stromal tissues, the formation of a placenta starts mainly in the trophectoderm layer. During early placentation, the human extravillous trophoblast (EVT) invades and remodels the maternal spiral arteries, causing the loss of arterial contractility. This EVT invasion is an essential process for embryo implantation and placental formation to support the subsequent placental function, maintaining adequate maternal blood flow into the intervillous spaces. The insufficient infiltration of EVT leads to placental dysfunction and induces various obstetrical disorders, such as preeclampsia [44]. In contrast to malignant cells, EVT invasion is confined spatially to the uterus and temporally to early pregnancy. Although various mechanisms for EVT invasion have been proposed, including growth factors [45, 46], the precise mechanism of how EVT invasion is induced toward maternal arteries or limited within one third of the uterine muscle layer is largely unknown.

We found that chemokine receptor/CCR1 was specifically and continuously expressed on EVT from the distal site of the cell column to the endovascular trophoblast through the shell. In the primary villous explant culture, CCR1 expression was induced on migrating EVT [47]. In addition, regulated on activation normal T cell expressed and secreted (RANTES), a ligand for CCR1, promoted the invasion of migrating EVT isolated from primary culture. From these findings, we proposed that chemokines are new regulators of EVT invasion toward maternal spiral arteries. An immunochemical study showed that RANTES was detected on decidual lymphocytes, while other CCR1 ligands, monocyte chemotactic protein-2 and macrophage inflammatory protein-1-α, were observed in decidual cells and EVT, respectively. Accordingly, we obtained no definite evidence about key cells to induce EVT invasion toward maternal spiral arteries [47]. Later, we found that CD41-positive platelets were deposited among endovascular trophoblasts and these platelets expressed P-selectin, an activation marker for platelets. Activated platelets secrete chemoattractive substances including RANTES. When EVT was cultured with platelets, the platelets attached to EVT and became activated, expressing P-selectin. In an invasion assay, platelets promoted EVT invasion partially through the CCR1 receptor system. Furthermore, during coculture with platelets for 48 h, EVT gradually became round or oval shaped, resembling the endovascular trophoblast. In addition, the expression of integrin α-1 was increased by platelets [48]. From these findings, we proposed that platelet-derived chemokines induce EVT invasion toward maternal spiral arteries and that platelets are also involved in EVT differentiation into endovascular trophoblasts. This system can explain the remodeling process of the maternal artery. The sequential accumulation and activation of platelets among EVT in the maternal artery make it possible for endovascular trophoblasts to gradually and proximally invade spiral arteries until the second trimester. Clinically, anticoagulation therapy using aspirin or heparin is well known to be effective for certain patients with habitual abortion or preeclampsia. Since these drugs can regulate the platelet function, we speculate that the above system is involved in the therapeutic mechanisms of aspirin and heparin.

4.3 Conclusion

In this chapter, we described a novel mechanism regarding the physiological roles of circulating blood cells in embryo-maternal cross talk and maternal tissue remodeling. PBMC play an important role in CL function, endometrial differentiation, and embryo invasion, while platelets are involved in maternal tissue-remodeling processes such as CL and placental formation. There is also cross talk between endocrine and immune systems, from PBMC to CL or from HCG to PBMC. When the endocrine mechanism does not adequately operate in infertile patients, we can apply PBMC therapy based on alternative mechanisms of the immune system. Clarification of the precise mechanism of how the maternal immune system recognizes the developing embryo will contribute to developing more effective treatment for infertility therapy.

References

1. Simón C, Pellicer A, Polan ML. Interleukin-1 system crosstalk between embryo and endometrium in implantation. Hum Reprod. 1995;10 Suppl 2:43–54. Review.
2. Lea RG, Sandra O. Immunoendocrine aspects of endometrial function and implantation. Reproduction. 2007;134:389–404. Review.
3. Granot I, Gnainsky Y, Dekel N. Endometrial inflammation and effect on implantation improvement and pregnancy outcome. Reproduction. 2012;144:661–8. doi:10.1530/REP-12-0217. Epub 2012 Oct 1. Review.
4. Schjenken JE, Robertson SA. Seminal fluid and immune adaptation for pregnancy – comparative biology in mammalian species. Reprod Domest Anim. 2014;49 Suppl 3:27–36. doi:10.1111/rda.12383. Review.
5. Espey LL. Current status of the hypothesis that mammalian ovulation is comparable to an inflammatory reaction. Biol Reprod. 1994;2:233–8. Review.
6. Fujiwara H, Maeda M, Imai K, Fukuoka M, Yasuda K, Horie K, Takakura K, Taii S, Mori T. Differential expression of aminopeptidase-N on human ovarian granulosa and theca cells. J Clin Endocrinol Metab. 1992;74:91–5.
7. Corner GW. The histological dating of the human corpus luteum of menstruation. Am J Anat. 1956;98:377–401.
8. Yan Z, Weich HA, Bernart W, Breckwoldt M, Neulen J. Vascular endothelial growth factor (VEGF) messenger ribonucleic acid (mRNA) expression in luteinized human granulosa cells in vitro. J Clin Endocrinol Metab. 1993;77:1723–5.
9. Koga K, Osuga Y, Tsutsumi O, Momoeda M, Suenaga A, Kugu K, Fujiwara T, Takai Y, Yano T, Taketani Y. Evidence for the presence of angiogenin in human follicular fluid and the up-regulation of its production by human chorionic gonadotropin and hypoxia. J Clin Endocrinol Metab. 2000;85:3352–5.
10. Fraser HM, Bell J, Wilson H, Taylor PD, Morgan K, Anderson RA, Duncan WC. Localization and quantification of cyclic changes in the expression of endocrine gland vascular endothelial growth factor in the human corpus luteum. J Clin Endocrinol Metab. 2005;90:427–34.
11. Sugino N, Suzuki T, Sakata A, Miwa I, Asada H, Taketani T, Yamagata Y, Tamura H. Angiogenesis in the human corpus luteum: changes in expression of angiopoietins in the corpus luteum throughout the menstrual cycle and in early pregnancy. J Clin Endocrinol Metab. 2005;90:6141–8.

12. Yen SSC. Endocrine-metabolic adaptations in pregnancy. In: Yen SSC, Jaffe RB, editors. Reproductive endocrinology. 3rd ed. Philadelphia: Saunders; 1991. p. 936–81.
13. Takao Y, Honda T, Ueda M, Hattori N, Yamada S, Maeda M, Fujiwara H, Mori T, Wimalasena J. Immunohistochemical localization of the LH/HCG receptor in human ovary: HCG enhances cell surface expression of LH/HCG receptor on luteinizing granulosa cells in vitro. Mol Hum Reprod. 1997;3:569–78.
14. Fraser HM, Duncan WC. Vascular morphogenesis in the primate ovary. Angiogenesis. 2005;8:101–16. Epub 2005 Oct 21. Review.
15. Emi N, Kanzaki H, Yoshida M, Takakura K, Kariya M, Okamoto N, Imai K, Mori T. Lymphocytes stimulate progesterone production by cultured human granulosa luteal cells. Am J Obstet Gynecol. 1991;165:1469–74.
16. Fukuoka M, Yasuda K, Emi N, Fujiwara H, Iwai M, Takakura K, Kanzaki H, Mori T. Cytokine modulation of progesterone and estradiol secretion in cultures of luteinized human granulosa cells. J Clin Endocrinol Metab. 1992;75:254–8.
17. Xie S, Luca M, Huang S, Gutman M, Reich R, Johnson JP, Bar-Eli M. Expression of MCAM/MUC18 by human melanoma cells leads to increased tumor growth and metastasis. Cancer Res. 1997;57:2295–303.
18. Egawa M, Yoshioka S, Higuchi T, Sato Y, Tatsumi K, Fujiwara H, Fujii S. Ephrin B1 is expressed on human luteinizing granulosa cells in corpora lutea of the early luteal phase: the possible involvement of the B class Eph-ephrin system during corpus luteum formation. J Clin Endocrinol Metab. 2003;88:4384–92.
19. Yoshioka S, Fujiwara H, Higuchi T, Yamada S, Maeda M, Fujii S. Melanoma cell adhesion molecule (MCAM/CD146) is expressed on human luteinizing granulosa cells: enhancement of its expression by hCG, interleukin-1 and tumour necrosis factor-α. Mol Hum Reprod. 2003;9:311–9.
20. Furukawa K, Fujiwara H, Sato Y, Zeng BX, Fujii H, Yoshioka S, Nishi E, Nishio T. Platelets are novel regulators of neovascularization and luteinization during human corpus luteum formation. Endocrinology. 2007;148:3056–64. Epub 2007 Apr 19.
21. Alam V, Altieri E, Zegers-Hochschild F. Preliminary results on the role of embryonic human chorionic gonadotrophin in corpus luteum rescue during early pregnancy and the relationship to abortion and ectopic pregnancy. Hum Reprod. 1999;14:2375–8.
22. Kratzer PG, Taylor RN. Corpus luteum function in early pregnancies is primarily determined by the rate of change of human chorionic gonadotropin levels. Am J Obstet Gynecol. 1990;163:1497–502.
23. Fujiwara H, Ueda M, Imai K, et al. Human leukocyte antigen-DR is a differentiation antigen for human granulosa cells. Biol Reprod. 1993;49:705–15.
24. Hattori N, Ueda M, Fujiwara H, Fukuoka M, Maeda M, Mori T. Human luteal cells express leukocyte functional antigen (LFA)-3. J Clin Endocrinol Metab. 1995;80:78–84.
25. Bukovsky A, Caudle MR, Carson RJ, Gaytán F, Huleihel M, Kruse A, Schatten H, Telleria CM. Immune physiology in tissue regeneration and aging, tumor growth, and regenerative medicine. Aging (Albany NY). 2009;1:157–81. Review.
26. Hashii K, Fujiwara H, Yoshioka S, et al. Peripheral blood mononuclear cells stimulate progesterone production by luteal cells derived from pregnant and non-pregnant women: possible involvement of interleukin-4 and interleukin-10 in corpus luteum function and differentiation. Hum Reprod. 1998;13:2738–44.
27. Fujiwara H. Hypothesis: immune cells contribute to systemic cross-talk between the embryo and mother during early pregnancy in cooperation with the endocrine system. Reprod Med Biol. 2006;5:19–29.
28. Psychoyos A. The implantation window: basic and clinical aspects. In: Mori T et al., editors. Perspectives in assisted reproduction. Rome: Ares Serono Symposia; 1993. p. 57–62.
29. Dey SK. Implantation. In: Adashi EY et al., editors. Reproductive endocrinology, surgery, and technology. Philadelphia: Lippincott–Raven; 1996. p. 421–34.

30. Takabatake K, Fujiwara H, Goto Y, et al. Intravenous administration of splenocytes in early pregnancy changes the implantation window in mice. Hum Reprod. 1997;12:583–5.
31. Takabatake K, Fujiwara H, Goto Y, et al. Splenocytes in early pregnancy promote embryo implantation by regulating endometrial differentiation in mice. Hum Reprod. 1997;12:2102–7.
32. Bhatt H, Brunet LJ, Stewart CL. Uterine expression of leukemia inhibitory factor coincides with the onset of blastocyst implantation. Proc Natl Acad Sci U S A. 1991;88:11408–12.
33. Fujita K, Nakayama T, Takabatake K, Higuchi T, Fujita J, Maeda M, Fujiwara H, Mori T. Administration of thymocytes derived from non-pregnant mice induces an endometrial receptive stage and leukaemia inhibitory factor expression in the uterus. Hum Reprod. 1998;13:2888–94.
34. Kosaka K, Fujiwara H, Tatsumi K, et al. Human peripheral blood mononuclear cells enhance cell-cell interaction between human endometrial epithelial cells and BeWo-cell spheroids. Hum Reprod. 2003;18:19–25.
35. Yoshioka S, Fujiwara H, Nakayama T, Kosaka K, Mori T, Fujii S. Intrauterine administration of autologous peripheral blood mononuclear cells promotes implantation rates in patients with repeated failure of IVF-embryo transfer. Hum Reprod. 2006;21:3290–4.
36. Nakayama T, Fujiwara H, Maeda M, Inoue T, Yoshioka S, Mori T, Fujii S. Human peripheral blood mononuclear cells (PBMC) in early pregnancy promote embryo invasion in vitro: HCG enhances the effects of PBMC. Hum Reprod. 2002;17:207–12.
37. Egawa H, Fujiwara H, Hirano T, Nakayama T, Higuchi T, Tatsumi K, Mori T, Fujii S. Peripheral blood mononuclear cells in early pregnancy promote invasion of human choriocarcinoma cell line, BeWo cells. Hum Reprod. 2002;17:473–80.
38. Adcock 3rd E, Teasdale T, August CS, Cox S, Meschia G, Ballaglia TC, Naughton MA. Human chorionic gonadotropin: its possible role in maternal lymphocyte suppression. Science. 1973;181:845–7.
39. Muchmore AV, Blaese RM. Immunoregulatory properties of fractions from human pregnancy urine: evidence that human chorionic gonadotropin is not responsible. J Immunol. 1977;118:881–6.
40. Kosaka K, Fujiwara H, Tatsumi K, et al. Human chorionic gonadotropin (HCG) activates monocytes to produce interleukin-8 via a different pathway from luteinizing hormone/HCG receptor system. J Clin Endocrinol Metab. 2002;87:5199–208.
41. Cole LA. Hyperglycosylated hCG. Placenta. 2007;28:977–86.
42. Cole LA. New discoveries on the biology and detection of human chorionic gonadotropin. Reprod Biol Endocrinol. 2009;7:8.
43. Handschuh K, Guibourdenche J, Tsatsaris V, Guesnon M, Laurendeau I, Evain-Brion D, Fournier T. Human chorionic gonadotropin produced by the invasive trophoblast but not the villous trophoblast promotes cell invasion and is down-regulated by peroxisome proliferator-activated receptor-γ. Endocrinology. 2007;148:5011–9.
44. Pijnenborg R, Bland JM, Robertson WB, Brosens I. Uteroplacental arterial changes related to interstitial trophoblast migration in early human pregnancy. Placenta. 1983;4:397–413.
45. Aplin JD. Implantation, trophoblast differentiation and haemochorial placentation: mechanistic evidence *in vivo* and *in vitro*. J Cell Sci. 1991;99:681–92.
46. Lala PK, Chakraborty C. Factors regulating trophoblast migration and invasiveness: possible derangements contributing to pre-eclampsia and fetal injury. Placenta. 2003;24:575–87.
47. Sato Y, Higuchi H, Yoshioka S, Tatsumi K, Fujiwara H, Fujii S. Trophoblasts acquire a chemokine receptor, CCR1, as they differentiate towards invasive phenotype. Development. 2003;130:5519–32.
48. Sato Y, Fujiwara H, Zeng B-X, Higuchi T, Yoshioka S, Fujii S. Platelet-derived soluble factors induce human extravillous trophoblast migration and differentiation: platelets are a possible regulator of trophoblast infiltration into maternal spiral arteries. Blood. 2005;106:428–35.

Chapter 5
Functional Role of Uterine Natural Killer Cells

Atsushi Fukui, Ayano Funamizu, Kohei Fuchinoue, Mai Kamoi, Ayako Taima, Rie Fukuhara, and Hideki Mizunuma

Abstract The most abundant cells in the uterine endometrium (from the secretory phase to the stage of early-pregnancy decidua) are natural killer (NK) cells. Endometrial (uterine) NK cells and decidual NK cells are phenotypically and functionally different from peripheral blood NK cells.

Endometrial and decidual NK cells express various activating and inhibitory receptors on the surface and can produce various cytokines. The number and proportion of these NK cells dramatically increase in the secretory phase and early pregnancy. Appropriate regulation of the number and proportion of NK cells leads to reproductive success. Dysregulation of these NK cells has been associated with problems related to reproductive immunology, such as recurrent pregnancy loss, implantation failure, and preeclampsia. That is, aberrant expression of surface markers, i.e., a higher percentage of cytotoxic $CD16^{+}/CD56^{dim}$ cells and a lower percentage of $NKp46^{+}$ NK cells, can cause reproductive failure. Moreover, there is anomalous spiral artery remodeling because of irregular production of NK cell cytokines such as interferon γ. This phenomenon may also cause reproductive failure.

Recently, a new type of NK cells, NK22 cells, was reported. NK22 cells are present in the uterine endometrium and decidua and may play some roles not only in the mucosal barrier but also in reproduction.

This chapter discusses variations in the expression of activating and inhibitory receptors on the surface of NK cells, production of cytokines and other angiogenic factors by NK cells, and involvement of NK22 cells in healthy and pathological reproduction.

Keywords NK cell • Natural cytotoxicity receptor • Cytokine • NK22 • Recurrent pregnancy loss

---

A. Fukui (✉) • A. Funamizu • K. Fuchinoue • M. Kamoi • A. Taima • R. Fukuhara • H. Mizunuma
Department of Obstetrics and Gynecology, Hirosaki University Graduate School of Medicine, 5 Zaifu-cho, Hirosaki, Aomori 036-8562, Japan
e-mail: a.fukuipon@mac.com

H. Kanzaki (ed.), *Uterine Endometrial Function*,
DOI 10.1007/978-4-431-55972-6_5

5.1 Introduction

Natural killer (NK) cells may play an important role in pregnancy; systemic regulation and local regulation of NK cells contribute to reproductive success. These cells express the specific surface marker CD56 and can be distinguished from other immune cells. NK cells constitute ~10 % of peripheral blood lymphocytes and 20–30 % of proliferative-phase uterine endometrial lymphocytes. In the secretory phase of the menstrual cycle and early-pregnancy decidua, NK cells constitute approximately 60–80 % of uterine endometrial lymphocytes as the blood progesterone level increases. As this level drops from the secretory phase to menstruation, NK cells die [1]. NK cells ($CD56^{+}$ cells) can be subdivided into $CD56^{bright}$ cells and $CD56^{dim}$ cells according to the intensity of CD56 fluorescent staining. $CD56^{bright}$ cells constitute 10 % of NK cells and represent the main population of uterine NK (uNK) and decidual NK (dNK) cells. The main function of $CD56^{bright}$ NK cells is production of cytokines such as interferon (IFN) γ and tumor necrosis factor (TNF) α. On the other hand, 90 % of NK cells are $CD56^{dim}$ cells, and the latter represent the main population of peripheral blood NK (pNK) cells. The main function of $CD56^{dim}$ cells is cytotoxicity.

NK cells express various activating and inhibitory receptors, and NK cell cytotoxicity is determined by the balance of these activating and inhibitory receptors. Namely, if a target cell binds to an NK cell's activating receptor, the NK cell can attack this target cell. Conversely, if the target cell binds to the NK cell's inhibitory receptor, then the NK cell does not attack the target cell. If the target cell binds to both the activating receptor and the inhibitory receptor on the NK cell, then the attack of the NK cell on the target cell will depend on the balance of expression of the activating and inhibitory receptors on the surface of the NK cell. Moreover, NK cells preferentially kill target cells with lower-than-normal expression of major histocompatibility complex (MHC) class I proteins, because in this case, fewer inhibitory receptors engage ligands. This concept is known as the *missing-self hypothesis* [2].

Regulation of gene expression in uNK, dNK, and pNK cells seems to be associated with various problems related to reproductive immunology, such as recurrent pregnancy loss (RPL), implantation failure, and preeclampsia. Because NK cells exist in the endometrium and decidua, it is possible that endometrial NK (eNK) cells or dNK cells perform a function in the establishment and maintenance of pregnancy.

At the implantation site, the chorion consists of syncytiotrophoblasts and cytotrophoblasts. These cells do not express classical class I human leukocyte antigen (HLA) A and HLA-B or class II (HLA-DP, HLA-DQ, or HLA-DR) alloantigens. As a consequence, syncytiotrophoblasts are vulnerable to the cytotoxicity of pNK cells. Therefore, both dNK (eNK) and pNK cells may be important for successful pregnancy.

5.2 Is It Possible to Predict a Pregnancy Outcome by NK Cells?

Recently, the predictive value of preconceptional activity of pNK cells was evaluated by Katano et al. [3]; these authors reported that quantification of pNK cells is not useful for evaluation of RPL. On the other hand, 22 studies on NK cells in female infertility and RPL were evaluated in a meta-analysis [4]. This meta-analysis evaluated the percentage of pNK and uNK cells in infertile women versus fertile women and showed no significant difference between the two groups. There are, however, significantly greater NK cell numbers in infertile women than in fertile controls [4]. Moreover, the percentage and number of pNK cells are significantly greater in women with RPL than in women without RPL [4]. Nevertheless, there is no significant difference in uNK cells between women with and without RPL [4]. Generally, uNK cells are considered more important than pNK cells in terms of reproductive health. Those authors concluded that the immune system is too complex, and one variable, such as the NK cell number, cannot predict outcomes in women with either infertility or RPL. A similar prognostic value of eNK and pNK cells was stated in another review [5]. The authors of the latter review concluded that a higher percentage of prepregnancy pNK cells and a greater number of prepregnancy-associated uNK cells are not associated with subsequent pregnancy outcomes in women with infertility or RPL. They admitted, however, that the utility of measuring the NK cell activity or number as a prognostic indicator of pregnancy success was still not known. On the other hand, various reports have shown usefulness of analysis of prepregnancy-associated pNK or eNK cells as an indicator of reproductive success [6–19]. For example, the number of prepregnancy-associated peripheral $CD56^{+}$ cells is greater in women with RPL [6–8], and the number of prepregnancy-associated $CD16^{+}/CD56^{dim}$ eNK cells is also significantly greater in women who miscarry [9] or those with RPL [18]. In women who experienced reproductive failure, expression of activating and inhibitory receptors on NK cells is altered [10, 12–14, 17]. It was reported that a high level of preconceptional NK cell activity (>46 %) and percentage (>16.4 %) predict spontaneous abortion and biochemical pregnancy [6, 19]. Similarly, the percentage of $CD56^{+}$ pNK cells in nonpregnant women with RPL [7] and in pregnant women [8] is significantly higher in comparison with that in nonpregnant or pregnant controls, respectively. As for the uterine endometrium, the percentage of prepregnancy-associated uterine $CD56^{bright}$ eNK cells is significantly lower in women with RPL than in healthy fertile nonpregnant women [18]. Moreover, the percentage of prepregnancy-associated uterine $CD16^{-}/CD56^{bright}$ eNK cells is significantly lower, whereas that of $CD16^{+}/CD56^{dim}$ NK cells is significantly higher in women with RPL than in healthy fertile nonpregnant women [18]. More studies are needed to determine definitively whether analysis of uNK cells and pNK cells is effective in prediction of pregnancy outcomes.

5.3 Where Do uNK Cells Come From?

As mentioned above, pNK cells constitute approximately 10–15 % of peripheral blood lymphocytes, and most of pNK cells are $CD16^{+}/CD56^{dim}$. On the other hand, uNK cells constitute approximately 20–40 % of uterine endometrial lymphocytes in the proliferative phase of the menstrual cycle, but 60–70 % in the secretory phase of the menstrual cycle, and 70–80 % in the first trimester of pregnancy. Most uNK cells are $CD16^{-}/CD56^{bright}$. That is to say, pNK cells and uNK cells are phenotypically different.

Where do uNK cells come from? How does the percentage or number of uNK cells among other cells increase from the proliferative phase to secretory phase? Is it different in the uterus? There are many questions about uNK cells, and the mechanisms underlying the changes in uNK cells have not been elucidated yet.

There are two hypotheses about the origin of uNK cells. One is in situ proliferation of residual cells, and the other is selective recruitment from circulating pNK cells [20]. Although uNK cells do not express progesterone receptors [21], progesterone may promote accumulation and differentiation of uNK cells indirectly. Cytokines such as interleukin (IL) 2, IL-12, and IL-15 are more powerful stimulators of proliferation of uNK cells [20]. IL-2, however, is not detectable in the healthy endometrium and decidua [22]. IL-12 can stimulate proliferation of NK cells but also enhances their cytolytic activity [23]. Thus, these cytokines do not seem to be prominent candidates as stimulators of the proliferation of uNK cells. Nevertheless, IL-15 stimulates proliferation of $CD16^{-}/CD56^{bright}$ eNK cells without increasing their cytolytic activity [24]. Therefore, IL-15 seems to be a stimulator of the proliferation of uNK cells, according to the data published to date.

On the other hand, Kitaya et al. clearly demonstrated the mechanisms of selective recruitment of $CD16^{-}/CD56^{bright}$ NK cells from circulating pNK cells via uterine endometrial vessels [20]. There are four steps of recruitment: tethering/rolling, activation, tight adhesion, and transendothelial migration. The first step of the recruitment is weak attachment of leukocytes to the endothelial cells. This process is mediated by the interaction between L-selectin on leukocytes and its ligand on endothelial cells. Next, the leukocytes are activated by chemokines produced by the endothelial cells. Then, the leukocytes tightly adhere to the endothelial cells. Finally, the leukocytes pass through the endothelial cell layers. Moreover, indirect evidence exists that $CD16^{-}/CD56^{bright}$ uNK cells originate from the $CD16^{-}/CD56^{bright}$ pNK cells [25, 26]. pNK cells and uNK cells express CXCR3 and CXCR4 chemokine receptors, but only CXCR4 is involved in migration of $CD16^{-}/CD56^{bright}$ NK cells. CXCL12 (ligand of CXCL4) is expressed in the trophoblasts, including those in spiral arteries [25]. Some dNK cells express CXCR3 and CXCR4, but others do not. Thus, some dNK cells may originate from local progenitor cells [25]. Transforming growth factor (TGF) β may convert $CD16^{+}/CD56^{dim}$ NK cells into $CD16^{-}/CD56^{bright}$ NK cells [27].

5.4 Activating and Inhibitory Receptors on NK Cells

5.4.1 The Activating Receptors on NK Cells

An NK cell expresses various kinds of activating and inhibitory receptors (Table 5.1). As for the activating receptors, studies have been performed on the expression levels of killer immunoglobulin-like receptors (KIR: KIR2DS1, KIR2DS2, KIR2DL4, and others), c-type lectin receptors (CD94/NKG2C and

Table 5.1 NK cell receptors

NK cell-activating receptors		NK cell inhibitory receptors	
Receptors	In reproductive failure (RPL)	Receptors	In reproductive failure (RPL)
MHC class I specific		*MHC class I specific*	
KIR (activating)		*KIR (inhibitory)*	
KIR2DS1 (CD158h)	→ (61)	KIR2DL1 (CD158a)	↓ (55–57, 60), → (61)
KIR2DS2 (CD158j)	→ (61)	KIR2DL2 (CD158b1)	↓ (58–60)
KIR2DL4 (CD158d)	→ (62)	KIR2DL3 (CD158b2)	↓ (58–60)
KIR2DS4 (CD158i),,	→ (61)	KIR2DL5	→ (61)
KIR2DS5 (CD158f)	→ (61)	KIR3DL1 (CD158e1)	?
KIR3DS1 (CD158e2)	?	KIR3DL2	Related with pre-eclampsia (63)
C-type lectin receptors		KIR3DL7	?
CD94/ NKG2C	→ (59)	*C-type lectin receptors*	
CD94/ NKG2E/H	?	CD94/NKG2A (CD159a)	→ (59)
Non MHC class I specific		CD161	↑ (60)
Natural cytotoxicity receptors (NCR)		*Immunoglobulin-like transcripts*	
NKp46	↑ (44, 45), ↓ (10–15, 42, 47, 48)	ILT-2	?
NKp44	↑ (44)		
NKp30	↑ (45)		
C-type lectin receptors			
NKG2D	→ (44)		
Others			
CD16	↑ (9, 18, 54)		

CD94/NKG2D), natural cytotoxicity receptors (NCRs: NKp46, NKp44, and NKp30), and CD16 in healthy pregnancy and in aberrant pregnancy, for example, in RPL and implantation failure.

5.4.1.1 KIRs

dNK cells recognize HLA-C expressed by the extravillous trophoblast (EVT; paternal HLA-C). These NK cells express KIRs, and the dominant ligand of KIRs is HLA-C. There are two types of KIR: one is an inhibitory receptor, and the other is an activating receptor. HLA-C can be classified into two functional groups by the presence of asparagine or lysine at position 80 in the amino acid sequence [28]. The former is HLA-C1, and the latter is HLA-C2. HLA-C2 is recognized by inhibitory KIR2DL1 (CD158a) and activating KIR2DS1 (CD158h), whereas HLA-C1 is recognized by inhibitory KIR2DL2 (CD158b1) and KIR2DL3 (CD158b2). HLA-C1 is not recognized by activating KIRs. Furthermore, HLA-C2 interacts with KIRs more strongly than does HLA-C1 [29]. KIRs can be subdivided into two functional groups—KIR A and KIR B—by the presence or absence of particular subsets of KIR genes. uNK cells can produce angiogenic factors and cytokines that induce invasion of the trophoblast. This process is enhanced by binding of a ligand to stimulatory KIR B and is reduced by binding of a ligand to KIR A [30]. Possible maternal KIR genotypes are AA (nonactivating receptor) or AB/BB (one or more activating receptors) [31]. Therefore, the combination of fetal HLA-C2 with maternal KIR B/B is sufficient for placentation and prevention of preeclampsia. On the other hand, the combination of fetal HLA-C2 with maternal KIR A/A poses a high risk of preeclampsia [32]. Similarly, the KIR A/A genotype is significantly more frequent among women with RPL, such as in women with preeclampsia [33]. The frequency of KIR2DL4 (CD158d) genes associated with the B haplotype is significantly reduced in women with RPL [33].

5.4.1.2 NCRs

NCRs include NKp46, NKp44, and NKp30 and are the major receptors involved in NK cell cytotoxicity. NCRs play a role in the recognition and lysis of tumor cells by NK cells. NKp46 and NKp44, but not NKp30, recognize viral proteins such as hemagglutinin of the influenza virus or hemagglutinin-neuraminidase of the parainfluenza virus [34, 35]. The endogenous cellular ligands recognized by NCRs have not been characterized yet. Recently, NCR ligands were found to be expressed by murine lymphoma and myeloma cell lines [36] and in human primary nevi and melanomas [37, 38]. The receptors NKp30 and NKp46 are expressed on the surface of activated and unactivated NK cells, whereas NKp44 is expressed only on the surface of activated NK cells. In addition, NKp30 and NKp46 perform functions in the cytotoxic activity and cytokine production of NK cells. Recently, we reported the role of the NKp46 in cytokine production by NK cells in an in vitro study on the peripheral blood and endometrium of infertile women [39]. We found that NKp46 expression is associated with cytokine-producing NK cells of both the CD56^{bright} and CD56^{dim} types (Fig. 5.1). Briefly, we demonstrated a direct relation between the expression of NKp46 on NK cells and the corresponding cytokine production [39]. Generally, NKp46^{+} NK cells can be subdivided into NKp46^{bright} cells and NKp46^{dim} cells. NKp46^{bright} NK cells in peripheral blood and in the uterine endometrium show significantly higher IFN-γ production than do NKp46^{dim} cells. Moreover, among NKp46^{+} pNK and uNK cells, secretion of IFN-γ is greatest in NKp46^{bright}/CD56^{bright} NK cells (Fig. 5.1). We concluded that NKp46 is involved in cytokine production by CD56^{+} NK cells in peripheral blood and in the uterine endometrium.

In addition, there is a recent report showing a relation between the expression of NKp46 on NK cells and production of cytokines such as IFN-γ by NK cells [40]. These researchers compared the production of cytokines by NK cells between NKp46^{+} NK cells and NK cells with weak expression of NKp46 (NKp46^{low} NK cells). The proportion of IFN-γ-producing NK cells in the spleen, lung, and mediastinal lymph nodes was significantly lower among NKp46^{low} NK cells than among NKp46^{+} NK cells. Moreover, the proportions of both IFN-γ-producing NK cells and IL-13-producing NK cells were significantly lower among NKp46^{low} NK cells than among NKp46^{+} NK cells.

It was reported that NKp46 is a ubiquitous marker of NK cells [41]. Our studies, however, have revealed that ~80 % of pNK cells and uterine endometrial CD56^{dim} cells express NKp46 and that 90 % of CD56^{bright} cells or more do so [10, 14]

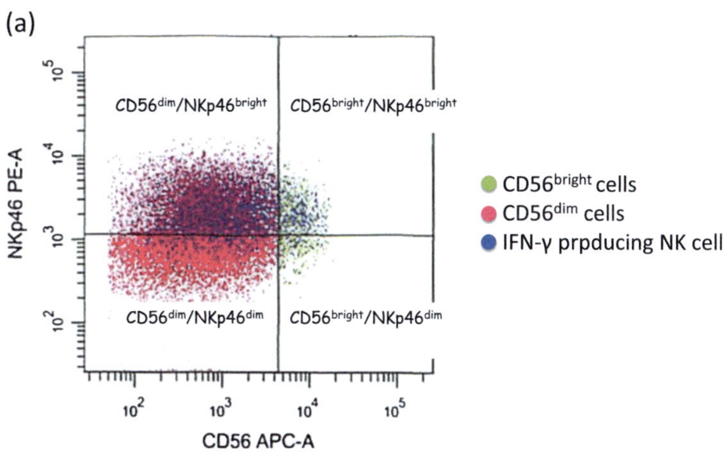

Fig. 5.1 The expression of NKp46 on NK cells and cytokine (IFN-γ) production by NK cells. (**a**) Representative dot plots of IFN-γ-producing NKp46^{+} NK cells. NKp46^{+} NK cells were collected using magnetic beads. Most of cytokine (IFN-γ)-producing NK cells are NKp46^{bright} NK cells. *Green dots*: CD56^{bright} NK cells. *Red dots*: CD56^{dim} NK cells. *Blue dots*: cytokine (IFN-γ)-producing NK cells. (**b**) IFN-γ-producing NKp46^{+} NK cells in peripheral blood. Secretion of IFN-γ is greatest in NKp46^{bright}/CD56^{bright} NK cells in peripheral blood. (**c**) IFN-γ-producing NKp46^{+} NK cells in endometrium. Secretion of IFN-γ is greatest in NKp46^{bright}/CD56^{bright} NK cells in uterine endometrium

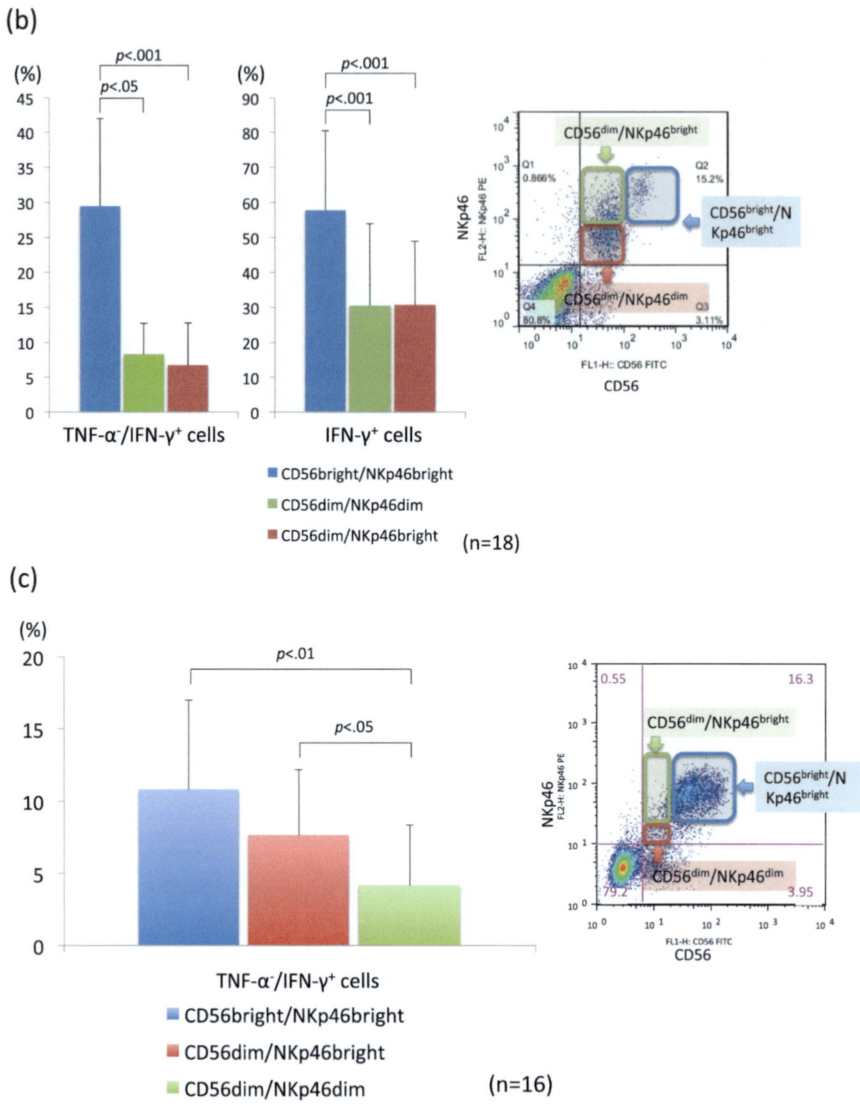

Fig. 5.1 (continued)

(Fig. 5.2). Moreover, the expression levels of NKp46 on pNK and/or eNK cells are low in women with various forms of reproductive failure such as RPL [10, 13–15], implantation failure, preeclampsia [42], and pelvic endometriosis [43].

5.4.1.3 NCRs in Reproduction

Various researchers evaluated the expression levels of NCRs on pNK or eNK cells in reproductive disorders such as RPL and implantation failure [10, 12–15, 39, 42–45].

As for the uterine endometrium, it was reported that eNK cells have a unique receptor repertoire, in particular, they express NKp46 and do not express (or express weakly) NKp30 and NKp44, both in the proliferative phase and in the secretory phase [46]. Nevertheless, our data showed that eNK cells express not only NKp46 but also NKp30 and NKp44 in the secretory phase [14, 15]. Our study showed that 80 % of eNK cells are NKp46$^+$, 25 % are NKp30$^+$, and ~10 % of pNK cells are NKp44$^+$ (Fig. 5.2).

The expression of NKp46, NKp44, NKp30, and NKG2D on dNK cells (CD56^{dim} cells and CD56^{bright} cells) was evaluated recently [44]. There was no significant difference between the frequency of NKG2D expression in women with RPL and in controls. On the other hand, the percentages of NKp46$^+$/CD56^{dim} NK cells and NKp44$^+$/CD56^{dim} NK cells were significantly higher in women with RPL than in controls. The frequency of NKp46 expression on CD16$^-$/CD56^{bright} cells was significantly greater than that on CD16$^+$/CD56^{dim} cells. Those authors concluded that enhanced cytotoxicity of dNK cells (because of stronger expression of NKp46 and NKp44 on these cells) may underlie the susceptibility to spontaneous abortion.

Recently, it was reported that the expression levels of NKp30 and NKp46 in peripheral blood and in placental tissue are significantly higher in women who experienced a spontaneous abortion than in women who undertook an elective abortion [45].

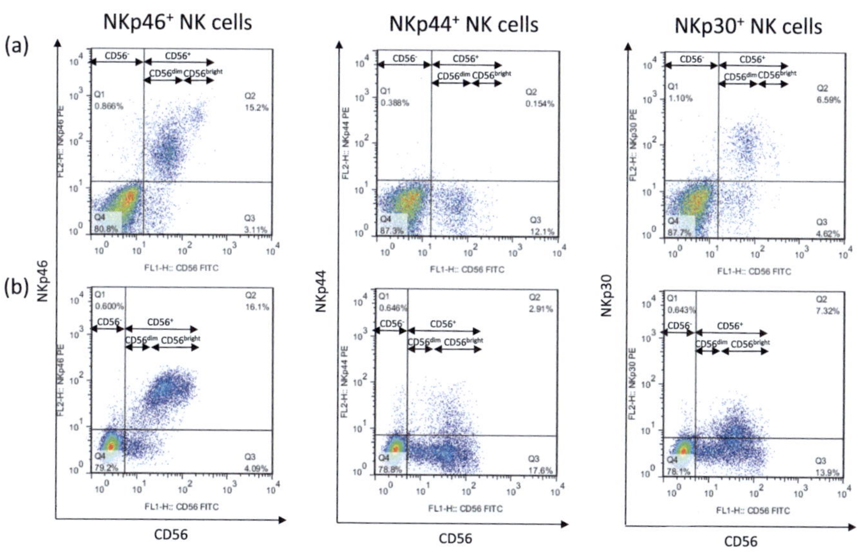

Fig. 5.2 Representative dot plots of the expression of natural cytotoxicity receptors (NKp46, NK44, and NKp30) on (**a**) peripheral blood NK cells and (**b**) endometrial NK cells

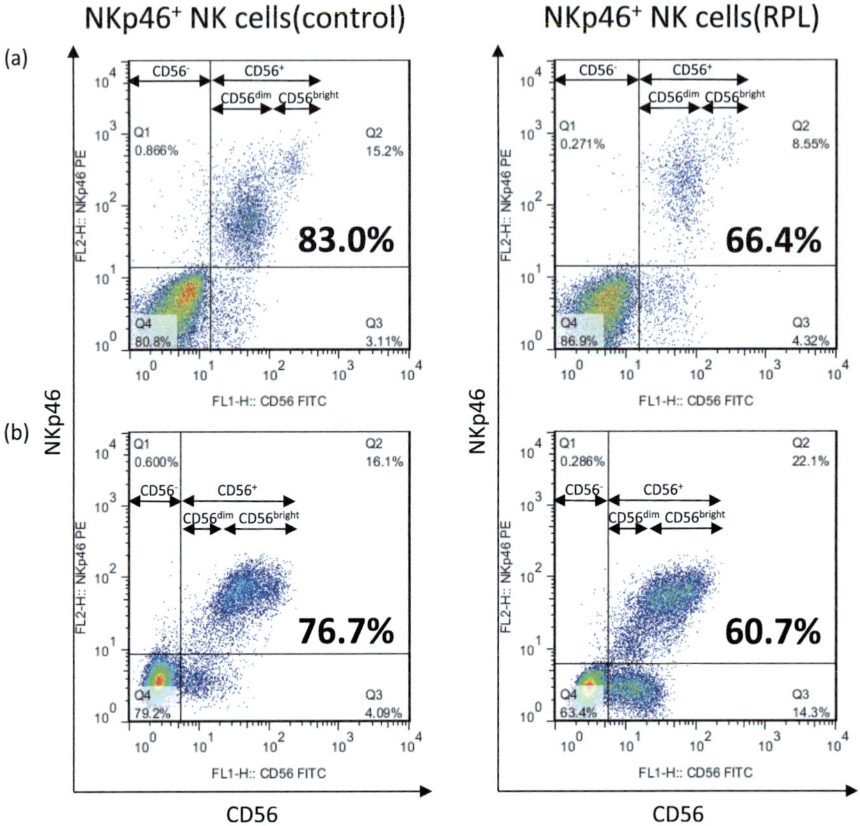

Fig. 5.3 Representative dot plots of the expression of NKp46 on peripheral blood (**a**) or endometrial (**b**) NK cells in woman with RPL or normal healthy woman. The expressions of NKp46 on NK cell both peripheral blood and endometrium are reduced in woman with RPL compared with normal healthy woman (control)

We also reported higher expression of NKp44 and NKp30 on some NK cells [15]. Because NCR$^+$ NK cells have stronger cytotoxicity, they may play a role in reproductive failure. On the other hand, in women with RPL or implantation failure, we detected lower expression of NKp46 on circulating pNK cells [10] (Fig. 5.3).

Briefly, the expression of CD56$^+$/NKp46$^+$ on NK cells was markedly lower in women with RPL than in healthy controls. Moreover, we showed that expression of NKp30 and NKp44 on CD56^{dim} NK cells is higher in aborted decidua than in the endometrium [14, 47]. The expression of NKp44 was significantly different between CD56^{dim} cells and CD56^{bright} cells in women with implantation failure; namely, the expression of NKp44 on CD56^{bright} cells was significantly upregulated in comparison with controls. This finding may mean that, in implantation failure, NK cell cytotoxicity is enhanced [47, 48].

5.4.1.4 CD16 on NK Cells

In peripheral blood, most of NK cells express CD16 (Fc receptor), and these cells show antibody-dependent cell-mediated cytotoxicity (ADCC). On the other hand, most eNK cells and dNK cells do not express CD16 (they are $CD16^{-}/CD56^{+}$) and have weaker cytotoxicity. The expression pattern $CD16^{+}/CD56^{dim}$ on cytotoxic eNK cells is more frequent in nonpregnant women with RPL than in healthy controls [18]. We also reported that cytotoxicity of pNK cells during the embryo transfer is significantly higher in women who will undergo miscarriage during the subsequent pregnancy than in women who will successfully deliver a baby [9]. Moreover, the percentage of peripheral blood $CD16^{+}/CD56^{+}$ cells during the embryo transfer is significantly higher in women who experience an in vitro fertilization and embryo transfer (IVF-ET) failure compared to women with IVF-ET success (implantation) [9]. The percentage of $CD16^{+}/CD56^{dim}$ eNK cells in the midsecretory phase immediately before IVF-ET is significantly higher in women who will miscarry during the subsequent pregnancy than in women who will deliver a baby successfully [9].

Human seminal plasma has powerful immunosuppressive properties because it contains high concentrations of various immunomodulators, such as prostaglandins (PGs), receptors for the Fc fragment of γ-globulin, and TGF-β [49]. In addition, seminal plasma and its principal constituent (PGs) stimulate the release of IL-8 and IL-10 from monocytes [50] and inhibit the release of IL-12 from blood cells [51]. It has been reported that TGF-β of seminal plasma initiates these immune responses and thereby induces type 2 maternal immune responses [52, 53]. Seminal plasma is considered a possible exogenous substance that affects the immune defenses of the female reproductive tract. Thus, appropriate sexual intercourse is important for successful pregnancy.

We evaluated the influence of sexual intercourse on uNK cells [54]. Namely, we evaluated the relation between the percentage of prepregnancy-associated uterine $CD16^{+}/CD56^{dim}$ eNK cells and sexual intercourse without any contraceptive devices. The results showed that when the sexual intercourse takes place before ovulation, the percentage of $CD16^{+}/CD56^{dim}$ uNK cells is significantly lower (whereas the percentage of $CD16^{-}/CD56^{bright}$ uNK cells is significantly higher) in comparison with postovulatory sexual intercourse and abstinence. Moreover, to clarify which is more important—the phase during which sexual intercourse occurs or the interval from sexual intercourse to the endometrial biopsy—we evaluated the correlation between the interval between sexual intercourse and the endometrial biopsy as well as the proportion of various uNK cell subpopulations. There was no correlation of $CD16^{-}/CD56^{bright}$ cells or $CD16^{+}/CD56^{dim}$ cells with the interval from the day of sexual intercourse to the day of the endometrial biopsy. According to these results, the timing of sexual intercourse as per the phase of the menstrual cycle, that is, sexual intercourse in the proliferative phase, is important for successful pregnancy. Therefore, the timing of sexual intercourse in the proliferative phase may facilitate the recruitment of NK cells into uterine endometrial tissue.

5.4.2 The Inhibitory Receptors on NK Cells

The expression of inhibitory receptors by NK cells, for example, KIRs (KIR2DL1 [CD158a], KIR2DL2/3 [CD158b], and others) and c-type lectin receptors (CD94/NKG2A), was evaluated during healthy pregnancy and during aberrant pregnancy, such as RPL and implantation failure [55–63].

As for KIR2DL1 (CD158a), the expression of CD158a and CD94 in NK ($CD56^{+}$) cells is significantly lower in miscarriage-associated decidua with a normal chromosomal karyotype of chorionic villi in comparison with miscarriage-associated decidua with an aberrant chromosomal karyotype of chorionic villi or decidua associated with an induced abortion [55]. The inhibitory receptor CD158a is expressed on NK cells not only in peripheral blood but also in the uterine endometrium and decidua [56]. Therefore, lower expression of CD158a on NK cells may cause excessive cytotoxicity of NK cells and thus may cause a miscarriage. We also characterized the expression of CD158a and CD158b on uNK cells [57]. Briefly, we found that there are significant positive correlations between these receptors and $CD56^{+}$ uNK cells. Moreover, there is a significant positive correlation between $CD16^{-}/CD56^{bright}$ cells and $CD158b^{+}$ NK cells [57]. There is a significant negative correlation between $CD16^{+}/CD56^{dim}$ cells and $CD158b^{+}$ NK cells [57]. It was reported that the number of NK cells expressing KIR2DL1, KIR2DL2, or KIR2DL3 (inhibitory KIR) is lower in aborters than in controls; some of these aborters have only one inhibitory KIR on their NK cells [58, 59]. Thus, women with RPL may have a limited number of inhibitory KIR on their NK cells. That is, some of them have only one inhibitory KIR, and most of them lack inhibitory KIRs [59]. In women with reproductive failure, such as miscarriage, the percentage of endometrial $CD16^{-}/CD56^{bright}$ cells is decreased and that of $CD16^{+}/CD56^{dim}$ cells is increased. These changes reduce expression of CD158b on eNK cells. The expression of KIRs on eNK cells may ensure the recognition of MHC class I molecules by maternal NK cells; therefore, the increase in the number of $CD16^{-}/CD56^{bright}$ cells may affect the maintenance of pregnancy. In peripheral blood, expression of CD158a and CD158b inhibitory receptors by $CD16^{+}/CD56^{dim}$ and $CD16^{-}/CD56^{bright}$ NK cells is significantly decreased, and expression of the CD161 activating receptor by $CD56^{+}/CD3^{+}$ natural killer T cells (NKT cells) is significantly increased in women with implantation failure in comparison with healthy controls [60]. Those researchers [60] concluded that there is an imbalance between expression levels of inhibitory and activating receptors in NK cells of women with implantation failure. This imbalance may explain the adverse reproductive outcomes.

5.5 Production of Cytokines and Angiogenic Factors by NK Cells

5.5.1 *Spiral Artery Remodeling by Cytokines and Angiogenic Factors Released by NK Cells*

NK cells, especially uNK and dNK cells, have one more very important function: cytokine production. NK cells can produce various cytokines, and the cytokine repertoire of NK cells consists mainly of type 1 cytokines such as IFN-γ and TNF-α. Nevertheless, NK cells can also produce other cytokines such as IL-4, IL-10, TGF-β, granulocyte macrophage colony-stimulating factor (GM-CSF), M-CSF, and leukemia inhibitory factor (LIF). Pregnancy is associated with a shift away from T helper 1 (Th1) immune responses and with a bias toward Th2 responses [64]. A similar concept has been demonstrated for NK cells, which can show comparable polarity (the NK1-NK2 concept) in their cytokine secretion profiles [11, 65–67].

uNK cells are a major source of uterine angiogenic growth factors, such as vascular endothelial growth factor (VEGF), angiopoietin 1 (Ang-1), Ang-2, and placental growth factor [68, 69], for the spiral artery modification. Thus, these uNK cells ensure sufficient dilatation of the maternal spiral arteries and the increased blood flow to the developing placenta [70]. EVTs invade decidua and remodel the uterine arteries by removing and replacing the vascular cells [71]; incomplete remodeling of the spiral artery causes reproductive failure such as RPL and preeclampsia [72]. Uterine spiral artery remodeling involves disorganization and clearance of the perivascular smooth muscle layer responsible for vasomotor control and clearance of the endothelial cells that line the walls of these arteries [73]. In spiral artery remodeling, there are four stages that are based on the extent of vascular smooth muscle cell (VSMC) disruption and loss [74]. Stage I corresponds to intact VSMC layers and endothelium without a detectable endovascular EVT. Stage II is disruption and a partial loss of VSMCs, the absence of an endovascular EVT, and minimal presence of the interstitial EVT. Stage III is major disorganization and clearance of VSMCs and the presence of an endovascular EVT. Stage IV corresponds to fully remodeled vessels with a complete loss of VSMCs and endothelium and replacement by an endovascular EVT. The main player of this event is uNK cells [75]. The initiation of the remodeling process is carried out primarily by uNK cells, and IFN-γ production by uNK cells is a key regulator of uterine artery remodeling [73, 76].

The main population of NK cells is IFN-γ- or TNF-α-producing NK1 cells. NK1 cytokines are important for maintenance of pregnancy [76–79] via angiogenesis and arterial remodeling as mentioned above. On the other hand, we have reported that there are significantly greater numbers of TNF-α- and/or IFN-γ-producing NK cells in women with reproductive failure such as RPL and implantation failure. It is likely that overproduction of these cytokines, especially TNF-α and IFN-γ, may be one of the causes of reproductive failure [11].

The role of TNF-α in pregnancy is quite complicated. Maternally derived TNF-α is essential for trophoblast cell growth, cell differentiation, and angiogenesis [78]. TNF-α can also regulate trophoblast apoptosis [79]. On the other hand, administration of high doses of TNF-α to pregnant mice results in an abortion [80]. The key for beneficial function of TNF-α in pregnancy seems to be its concentration. Moreover, we reported that even before pregnancy, there is dysregulation of cytokine production in NK cells and an NK1 shift in women with RPL or implantation failure [11]. It was reported that TNF-α expression in NK cells is downregulated when these cells encounter trophoblast cells. This result suggests that trophoblast cells may modulate the TNF-α secretion by NK cells [81]. Whether the dysregulation of cytokine production in women with RPL has any association with previous exposure of NK cells to trophoblast cells is a topic for future studies.

TNF-α- and IFN-γ-producing NK cells can be classified into three groups [11]: TNF-α-only-producing NK cells, IFN-γ-only-producing NK cells, and NK cells producing both TNF-α and IFN-γ. We reported that IFN-γ-only-producing NK cells are significantly downregulated in women with implantation failure in comparison with controls [11]. IFN-γ has been reported to play an essential role in implantation, especially in the vascular remodeling process. Genetic disruption of IFN-γ in uNK cells in mice results in the absence of the pregnancy-induced spiral artery modification. uNK cell-derived IFN-γ modifies the expression of genes in the uterine vasculature and stroma, and this change causes instability of the vessels and facilitates the pregnancy-induced remodeling of decidual arteries [76]. On the other hand, during healthy pregnancy, IFN-γ production by pNK cells is significantly reduced in comparison with the nonpregnant state [82]. A relative lack of INF-γ production by CD56^{bright} NK cells in women with implantation failure may explain the inefficiency of embryonic implantation.

We demonstrated that GM-CSF-producing NK cells tend to be downregulated in women with RPL or implantation failure in comparison with controls [11]. GM-CSF has been implicated in the morphological and functional development of the placenta [83]. GM-CSF promotes differentiation, DNA replication, and enhanced secretory activity of cytotrophoblast cells [83]. GM-CSF stimulates placental-cell proliferation and may cause the trophoblast to secrete human chorionic gonadotropin (hCG) and human placental lactogen (hPL). Anomalous functioning and development of the placenta have been described in GM-CSF-deficient mice [84]. GM-CSF is upregulated in healthy pregnancy and strongly downregulated in pregnant women with RPL [85].

5.5.2 The Expression of NCR on NK Cells and Production of Cytokines by NK Cells

As mentioned above, there is a direct relation between expression of NKp46 on NK cells and production of the corresponding cytokines [39]. There is a pathological correlation between NCR expression and cytokine production by pNK cells in reproductive failure such as RPL and implantation failure [13]. In brief, there is a significant positive correlation between the percentage of CD56^{bright}/NKp46^{+} cells and the percentage of TNF-α- and IFN-γ-producing NK cells in healthy women, whereas women with RPL show a significant negative correlation [13]. The presence of aberrant NCRs or inhibition of the related signal transduction after NCR activation may disturb the relation between NCRs and cytokine production in NK cells. In addition, excessive production of proinflammatory cytokines by NK cells may be mediated by disruption of NCR expression or of the related signal transduction.

5.5.3 NK22 Cells

A new type of NKp46^{+} NK cells that produces IL-22 has been reported [86–88]. These cells, known as NCR22 or NK22 cells, can be distinguished from conventional NK cells, and it is thought that their IL-22 production may be involved in mucosal immune defenses of the respiratory organs, intestines, skin, and liver. In these organs, NK22 cells perform an important function in the prevention of infection and protection of the mucosa, whereas aberrations in the NK22 cell function can cause asthma, ulcerative colitis, psoriasis, or atopic dermatitis [89]. In the liver, NK22 cells are involved in the proliferation of hepatocytes, and NK22 cell dysfunction is associated with hepatocellular carcinoma. It was also reported that IL-22-producing NK cells are present in the uterine mucosa [90]. Their function in reproduction is still unclear.

A reduced protein level of IL-22 or IL-22 receptor α1 (IL-22R1) in chorionic villi may be involved in spontaneous miscarriages [91]. The expression of IL-22R1 in the villi of women who experienced an unexplained spontaneous miscarriage is lower than that in women at the early stage of healthy pregnancy. Those researchers [91] concluded that IL-22 that is secreted by dNK cells may promote the survival of trophoblasts and participate in the maintenance of pregnancy by binding to IL-22R1 [91]. Therefore, IL-22-producing NK cells ought to be important for reproduction.

We have also evaluated the physiological role of NK22 cells in women with or without unexplained RPL by means of samples of the prepregnancy-associated peripheral blood and midsecretory uterine endometrium [92]. The proportion of NK22 cells in peripheral blood and in the uterine endometrium is significantly higher in women with unexplained RPL than in women without RPL [92]. Moreover, there are significant negative correlations between the percentage of NK22 cells and the percentage of TNF-α- or IFN-γ-producing NK cells. That is, a higher proportion of IL-22-producing NK cells corresponds to lower production of IFN-γ and TNF-α both in peripheral blood and in the endometrium. This reduction in TNF-α and IFN-γ production by NK cells is seen only in women with unexplained RPL, not in women without RPL. As for other cytokines such as IL-4, IL-10, and TGF-β, there is no relation between NK cells producing them and NK22 cells.

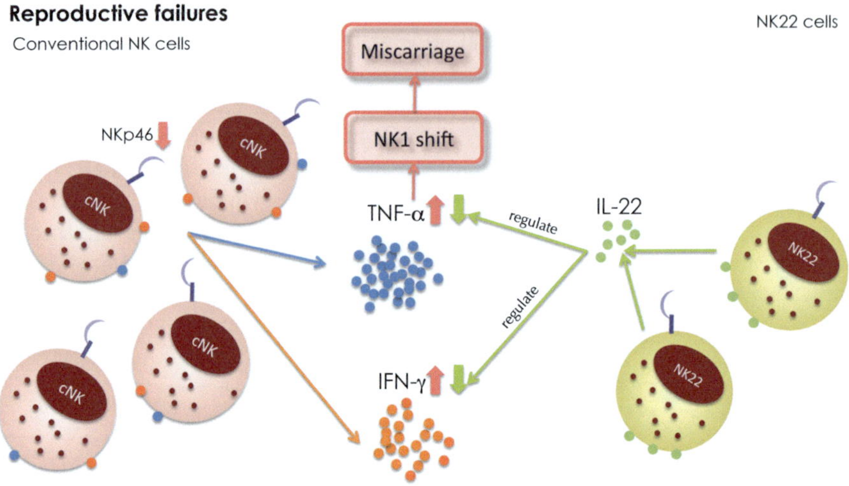

Fig. 5.4 Possible roles of natural cytotoxicity receptors and NK22 cells in reproduction. There is a lower expression of NKp46 on peripheral blood and uterine NK cells in women with reproductive failure such as recurrent pregnancy loss or implantation failure. It may lead to NK1 shift (increase of TNF-α and IFN-γ) of NK cells. Then, the proportion of NK22 cells increases in women with reproductive failure, and these cells may regulate the cytokine production by NK cells

In addition, there is a significant negative correlation between NK22 cells and NKp46$^+$ NK cells. These findings shed some light on the relations among NKp46$^+$ NK cells, production of cytokines (IFN-γ and TNF-α) by NK cells, and NK22 cells [92]. In healthy women (without RPL), NK22 cells may produce IL-22. Nevertheless, IFN-γ and TNF-α production by NK cells is normal; therefore, NK22 cells do not need to regulate the production of cytokines by conventional NK cells. On the other hand, in women with reproductive failure, expression of NKp46 is lowered and production of IFN-γ and TNF-α is enhanced. Subsequently, the proportion of NK22 cells increases, and these cells may regulate the production of cytokines by NK cells. Thus, NK22 cells may perform some regulatory function in women with reproductive failure (Fig. 5.4).

5.6 Conclusions

Appropriate expression of surface markers and orderly production of cytokines by NK cells lead to favorable pregnancy outcomes. It is a still controversial topic whether anomalous gene expression or aberrant function of NK cells (especially uterine eNK and dNK cells) causes RPL, implantation failure, or pregnancy-induced hypertension. It is obvious, however, that these disorders are associated with aberrant expression of activating and inhibitory surface markers such as CD16,

NCRs, and KIRs on NK cells. Moreover, production of cytokines by NK cells is anomalous in women with reproductive failure. Further research is needed to solve the puzzle of NK cells in women with reproductive failure.

References

1. Sharma S. Natural killer cells and regulatory T cells in early pregnancy loss. Int J Dev Biol. 2014;58(2–4):219–29.
2. Karre K, Ljunggren HG, Piontek G, Kiessling R. Selective rejection of H-2-deficient lymphoma variants suggests alternative immune defence strategy. Nature. 1986;319(6055):675–8.
3. Katano K, Suzuki S, Ozaki Y, Suzumori N, Kitaori T, Sugiura-Ogasawara M. Peripheral natural killer cell activity as a predictor of recurrent pregnancy loss: a large cohort study. Fertil Steril. 2013;100(6):1629–34.
4. Seshadri S, Sunkara SK. Natural killer cells in female infertility and recurrent miscarriage: a systematic review and meta-analysis. Hum Reprod Update. 2014;20(3):429–38.
5. Tang AW, Alfirevic Z, Quenby S. Natural killer cells and pregnancy outcomes in women with recurrent miscarriage and infertility: a systematic review. Hum Reprod. 2011;26(8):1971–80.
6. Kwak-Kim J, Gilman-Sachs A. Clinical implication of natural killer cells and reproduction. Am J Reprod Immunol. 2008;59(5):388–400.
7. Kwak JY, Beaman KD, Gilman-Sachs A, Ruiz JE, Schewitz D, Beer AE. Up-regulated expression of CD56+, CD56+/CD16+, and CD19+ cells in peripheral blood lymphocytes in pregnant women with recurrent pregnancy losses. Am J Reprod Immunol. 1995;34(2):93–9.
8. Coulam CB, Goodman C, Roussev RG, Thomason EJ, Beaman KD. Systemic CD56+ cells can predict pregnancy outcome. Am J Reprod Immunol. 1995;33(1):40–6.
9. Fukui A, Fujii S, Yamaguchi E, Kimura H, Sato S, Saito Y. Natural killer cell subpopulations and cytotoxicity for infertile patients undergoing in vitro fertilization. Am J Reprod Immunol. 1999;41(6):413–22.
10. Fukui A, Ntrivalas E, Gilman-Sachs A, Kwak-Kim J, Lee SK, Levine R, et al. Expression of natural cytotoxicity receptors and a2V-ATPase on peripheral blood NK cell subsets in women with recurrent spontaneous abortions and implantation failures. Am J Reprod Immunol. 2006;56(5–6):312–20.
11. Fukui A, Kwak-Kim J, Ntrivalas E, Gilman-Sachs A, Lee SK, Beaman K. Intracellular cytokine expression of peripheral blood natural killer cell subsets in women with recurrent spontaneous abortions and implantation failures. Fertil Steril. 2008;89(1):157–65.
12. Fukui A, Nakamura R, Yamada K, Fukuhara R, Kimura H, Fujii S, et al. Expression of natural cytotoxicity receptors and intracellular cytokine production of natural killer cell subsets in women with implantation failures. J Fertil Implant. 2009;26(1):341–7.
13. Fukui A, Ntrivalas E, Fukuhara R, Fujii S, Mizunuma H, Gilman-Sachs A, et al. Correlation between natural cytotoxicity receptors and intracellular cytokine expression of peripheral blood NK cells in women with recurrent pregnancy losses and implantation failures. Am J Reprod Immunol. 2009;62(6):371–80.
14. Fukui A, Nakamura R, Yamada K, Yokota M, Fukuhara R, Kimura H, et al. Expression of natural cytotoxicity receptors on midsecretory endometrial or decidual natural killer cells. J Fertil Implant. 2010;27(1):369–74.
15. Fukui A, Funamizu A, Yokota M, Yamada K, Nakamua R, Fukuhara R, et al. Uterine and circulating natural killer cells and their roles in women with recurrent pregnancy loss, implantation failure and preeclampsia. J Reprod Immunol. 2011;90(1):105–10.
16. Chernyshov VP, Sudoma IO, Dons'koi BV, Kostyuchyk AA, Masliy YV. Elevated NK cell cytotoxicity, CD158a expression in NK cells and activated T lymphocytes in peripheral blood of women with IVF failures. Am J Reprod Immunol. 2010;64(1):58–67.

17. Junovich G, Azpiroz A, Incera E, Ferrer C, Pasqualini A, Gutierrez G. Endometrial CD16(+) and CD16(−) NK cell count in fertility and unexplained infertility. Am J Reprod Immunol. 2013;70(3):182–9.
18. Lachapelle MH, Miron P, Hemmings R, Roy DC, Endometrial T. B, and NK cells in patients with recurrent spontaneous abortion. Altered profile and pregnancy outcome. J Immunol. 1996;156(10):4027–34.
19. Yamada H, Morikawa M, Kato EH, Shimada S, Kobashi G, Minakami H. Pre-conceptional natural killer cell activity and percentage as predictors of biochemical pregnancy and spontaneous abortion with normal chromosome karyotype. Am J Reprod Immunol. 2003;50 (4):351–4.
20. Kitaya K, Yamaguchi T, Yasuo T, Okubo T, Honjo H. Post-ovulatory rise of endometrial CD16(−) natural killer cells: in situ proliferation of residual cells or selective recruitment from circulating peripheral blood? J Reprod Immunol. 2007;76(1–2):45–53.
21. Henderson TA, Saunders PT, Moffett-King A, Groome NP, Critchley HO. Steroid receptor expression in uterine natural killer cells. J Clin Endocrinol Metab. 2003;88(1):440–9.
22. Hamai Y, Fujii T, Yamashita T, Kozuma S, Okai T, Taketani Y. Pathogenetic implication of interleukin-2 expressed in preeclamptic decidual tissues: a possible mechanism of deranged vasculature of the placenta associated with preeclampsia. Am J Reprod Immunol. 1997;38 (2):83–8.
23. Hayakawa S, Nagai N, Kanaeda T, Karasaki-Suzuki M, Ishii M, Chishima F, et al. Interleukin-12 augments cytolytic activity of peripheral and decidual lymphocytes against choriocarcinoma cell lines and primary culture human placental trophoblasts. Am J Reprod Immunol. 1999;41(5):320–9.
24. Kitaya K, Yasuda J, Nakayama T, Fushiki S, Honjo H. Effect of female sex steroids on human endometrial CD16neg CD56bright natural killer cells. Fertil Steril. 2003;79 Suppl 1:730–4.
25. Hanna J, Wald O, Goldman-Wohl D, Prus D, Markel G, Gazit R, et al. CXCL12 expression by invasive trophoblasts induces the specific migration of CD16- human natural killer cells. Blood. 2003;102(5):1569–77.
26. Carlino C, Stabile H, Morrone S, Bulla R, Soriani A, Agostinis C, et al. Recruitment of circulating NK cells through decidual tissues: a possible mechanism controlling NK cell accumulation in the uterus during early pregnancy. Blood. 2008;111(6):3108–15.
27. Lee JY, Lee M, Lee SK. Role of endometrial immune cells in implantation. Clin Exp Reprod Med. 2011;38(3):119–25.
28. Mandelboim O, Reyburn HT, Vales-Gomez M, Pazmany L, Colonna M, Borsellino G, et al. Protection from lysis by natural killer cells of group 1 and 2 specificity is mediated by residue 80 in human histocompatibility leukocyte antigen C alleles and also occurs with empty major histocompatibility complex molecules. J Exp Med. 1996;184(3):913–22.
29. Rajagopalan S, Long EO. Understanding how combinations of HLA and KIR genes influence disease. J Exp Med. 2005;201(7):1025–9.
30. Moffett A, Hiby SE. How Does the maternal immune system contribute to the development of pre-eclampsia? Placenta. 2007;28(Suppl A):S51–6.
31. Redman CW, Sargent IL. Immunology of pre-eclampsia. Am J Reprod Immunol. 2010;63 (6):534–43.
32. Hiby SE, Walker JJ, O'Shaughnessy KM, Redman CW, Carrington M, Trowsdale J, et al. Combinations of maternal KIR and fetal HLA-C genes influence the risk of preeclampsia and reproductive success. J Exp Med. 2004;200(8):957–65.
33. Hiby SE, Regan L, Lo W, Farrell L, Carrington M, Moffett A. Association of maternal killer-cell immunoglobulin-like receptors and parental HLA-C genotypes with recurrent miscarriage. Hum Reprod. 2008;23(4):972–6.
34. Arnon TI, Lev M, Katz G, Chernobrov Y, Porgador A, Mandelboim O. Recognition of viral hemagglutinins by NKp44 but not by NKp30. Eur J Immunol. 2001;31(9):2680–9.

35. Mandelboim O, Lieberman N, Lev M, Paul L, Arnon TI, Bushkin Y, et al. Recognition of haemagglutinins on virus-infected cells by NKp46 activates lysis by human NK cells. Nature. 2001;409(6823):1055–60.
36. Halfteck GG, Elboim M, Gur C, Achdout H, Ghadially H, Mandelboim O. Enhanced in vivo growth of lymphoma tumors in the absence of the NK-activating receptor NKp46/NCR1. J Immunol. 2009;182(4):2221–30.
37. Lakshmikanth T, Burke S, Ali TH, Kimpfler S, Ursini F, Ruggeri L, et al. NCRs and DNAM-1 mediate NK cell recognition and lysis of human and mouse melanoma cell lines in vitro and in vivo. J Clin Invest. 2009;119(5):1251–63.
38. Cagnano E, Hershkovitz O, Zilka A, Bar-Ilan A, Golder A, Sion-Vardy N, et al. Expression of ligands to NKp46 in benign and malignant melanocytes. J Invest Dermatol. 2008;128 (4):972–9.
39. Yokota M, Fukui A, Funamizu A, Nakamura R, Kamoi M, Fuchinoue K, et al. Role of NKp46 expression in cytokine production by CD56-positive NK cells in the peripheral blood and the uterine endometrium. Am J Reprod Immunol. 2013;69(3):202–11.
40. Ghadially H, Horani A, Glasner A, Elboim M, Gazit R, Shoseyov D, et al. NKp46 regulates allergic responses. Eur J Immunol. 2013;43(11):3006–16.
41. Sivori S, Vitale M, Morelli L, Sanseverino L, Augugliaro R, Bottino C, et al. p46, a novel natural killer cell-specific surface molecule that mediates cell activation. J Exp Med. 1997;186 (7):1129–36.
42. Fukui A, Yokota M, Funamizu A, Nakamua R, Fukuhara R, Yamada K, et al. Changes of NK cells in preeclampsia. Am J Reprod Immunol. 2012;67(4):278–86.
43. Funamizu A, Fukui A, Kamoi M, Fuchinoue K, Yokota M, Fukuhara R, et al. Expression of natural cytotoxicity receptors on peritoneal fluid natural killer cell and cytokine production by peritoneal fluid natural killer cell in women with endometriosis. Am J Reprod Immunol. 2014;71(4):359–67.
44. Zhang Y, Zhao A, Wang X, Shi G, Jin H, Lin Q. Expressions of natural cytotoxicity receptors and NKG2D on decidual natural killer cells in patients having spontaneous abortions. Fertil Steril. 2008;90(5):1931–7.
45. Shemesh A, Tirosh D, Sheiner E, Tirosh NB, Brusilovsky M, Segev R, et al. First trimester pregnancy loss and the expression of alternatively spliced NKp30 isoforms in maternal blood and placental tissue. Front Immunol. 2015;6:189.
46. Manaster I, Mizrahi S, Goldman-Wohl D, Sela HY, Stern-Ginossar N, Lankry D, et al. Endometrial NK cells are special immature cells that await pregnancy. J Immunol. 2008;181(3):1869–76.
47. Fukui A. Uterine and circulating natural killer cells and their roles in women with recurrent pregnancy losses, implantation failures or preeclampsia. J Reprod Immunol. 2010;86(2):14.
48. Fukui A. NK cells and its role in reproduction. Am J Reprod Immunol. 2010;64(Supplement 1):1.
49. Nocera M, Chu TM. Transforming growth factor beta as an immunosuppressive protein in human seminal plasma. Am J Reprod Immunol. 1993;30(1):1–8.
50. Denison FC, Grant VE, Calder AA, Kelly RW. Seminal plasma components stimulate interleukin-8 and interleukin-10 release. Mol Hum Reprod. 1999;5(3):220–6.
51. Kelly RW, Carr GG, Critchley HO. A cytokine switch induced by human seminal plasma: an immune modulation with implications for sexually transmitted disease. Hum Reprod. 1997;12 (4):677–81.
52. Robertson SA, Mau VJ, Hudson SN, Tremellen KP. Cytokine-leukocyte networks and the establishment of pregnancy. Am J Reprod Immunol. 1997;37(6):438–42.
53. Tremellen KP, Seamark RF, Robertson SA. Seminal transforming growth factor beta1 stimulates granulocyte-macrophage colony-stimulating factor production and inflammatory cell recruitment in the murine uterus. Biol Reprod. 1998;58(5):1217–25.

54. Kimura H, Fukui A, Fujii S, Yamaguchi E, Kasai G, Mizunuma H. Timed sexual intercourse facilitates the recruitment of uterine CD56(bright) natural killer cells in women with infertility. Am J Reprod Immunol. 2009;62(2):118–24.
55. Yamada H, Shimada S, Morikawa M, Iwabuchi K, Kishi R, Onoe K, et al. Divergence of natural killer cell receptor and related molecule in the decidua from sporadic miscarriage with normal chromosome karyotype. Mol Hum Reprod. 2005;11(6):451–7.
56. Verma S, King A, Loke YW. Expression of killer cell inhibitory receptors on human uterine natural killer cells. Eur J Immunol. 1997;27(4):979–83.
57. Fukui A, Fujii S, Kasai G, Kimura H, Yamaguchi E, Mizunuma H. Correlation with endometrial NK cell, KIRs and NKT cells. IFFS 2001 Selected Communications. 2001:65–71.
58. Acar N, Ustunel I, Demir R. Uterine natural killer (uNK) cells and their missions during pregnancy: a review. Acta Histochem. 2011;113(2):82–91.
59. Varla-Leftherioti M, Spyropoulou-Vlachou M, Niokou D, Keramitsoglou T, Darlamitsou A, Tsekoura C, et al. Natural killer (NK) cell receptors' repertoire in couples with recurrent spontaneous abortions. Am J Reprod Immunol. 2003;49(3):183–91.
60. Ntrivalas EI, Bowser CR, Kwak-Kim J, Beaman KD, Gilman-Sachs A. Expression of killer immunoglobulin-like receptors on peripheral blood NK cell subsets of women with recurrent spontaneous abortions or implantation failures. Am J Reprod Immunol. 2005;53(5):215–21.
61. Hong Y, Wang X, Lu P, Song Y, Lin Q. Killer immunoglobulin-like receptor repertoire on uterine natural killer cell subsets in women with recurrent spontaneous abortions. Eur J Obstet Gynecol Reprod Biol. 2008;140(2):218–23.
62. Yan WH, Lin A, Chen BG, Zhou MY, Dai MZ, Chen XJ, et al. Possible roles of KIR2DL4 expression on uNK cells in human pregnancy. Am J Reprod Immunol. 2007;57(4):233–42.
63. Wang XL, Wang Q, Sun CJ, Zhang WY. Genetic polymorphisms of killer cell immunoglobulin-like receptor 3DL2 in preeclampsia. J Perinat Med. 2011;39(3):273–8.
64. Kwak-Kim JY, Gilman-Sachs A, Kim CE. T helper 1 and 2 immune responses in relationship to pregnancy, nonpregnancy, recurrent spontaneous abortions and infertility of repeated implantation failures. Chem Immunol Allergy. 2005;88:64–79.
65. Carter LL, Dutton RW. Relative perforin- and Fas-mediated lysis in T1 and T2 CD8 effector populations. J Immunol. 1995;155(3):1028–31.
66. Peritt D, Robertson S, Gri G, Showe L, Aste-Amezaga M, Trinchieri G. Differentiation of human NK cells into NK1 and NK2 subsets. J Immunol. 1998;161(11):5821–4.
67. Higuma-Myojo S, Sasaki Y, Miyazaki S, Sakai M, Siozaki A, Miwa N, et al. Cytokine profile of natural killer cells in early human pregnancy. Am J Reprod Immunol. 2005;54(1):21–9.
68. Lash GE, Schiessl B, Kirkley M, Innes BA, Cooper A, Searle RF, et al. Expression of angiogenic growth factors by uterine natural killer cells during early pregnancy. J Leukoc Biol. 2006;80(3):572–80.
69. Lash GE, Bulmer JN. Do uterine natural killer (uNK) cells contribute to female reproductive disorders? J Reprod Immunol. 2011;88(2):156–64.
70. Hanna J, Goldman-Wohl D, Hamani Y, Avraham I, Greenfield C, Natanson-Yaron S, et al. Decidual NK cells regulate key developmental processes at the human fetal-maternal interface. Nat Med. 2006;12(9):1065–74.
71. Pijnenborg R, Vercruysse L, Hanssens M. The uterine spiral arteries in human pregnancy: facts and controversies. Placenta. 2006;27(9–10):939–58.
72. Brosens IA, Robertson WB, Dixon HG. The role of the spiral arteries in the pathogenesis of preeclampsia. Obstet Gynecol Annu. 1972;1:177–91.
73. Tessier DR, Yockell-Lelievre J, Gruslin A. Uterine spiral artery remodeling: the role of uterine natural killer cells and extravillous trophoblasts in normal and high-risk human pregnancies. Am J Reprod Immunol. 2015;74(1):1–11.
74. Smith SD, Dunk CE, Aplin JD, Harris LK, Jones RL. Evidence for immune cell involvement in decidual spiral arteriole remodeling in early human pregnancy. Am J Pathol. 2009;174(5):1959–71.

75. Robson A, Harris LK, Innes BA, Lash GE, Aljunaidy MM, Aplin JD, et al. Uterine natural killer cells initiate spiral artery remodeling in human pregnancy. FASEB J. 2012;26 (12):4876–85.
76. Ashkar AA, Di Santo JP, Croy BA. Interferon gamma contributes to initiation of uterine vascular modification, decidual integrity, and uterine natural killer cell maturation during normal murine pregnancy. J Exp Med. 2000;192(2):259–70.
77. Bulmer JN, Lash GE. Human uterine natural killer cells: a reappraisal. Mol Immunol. 2005;42 (4):511–21.
78. Terranova PF, Hunter VJ, Roby KF, Hunt JS. Tumor necrosis factor-alpha in the female reproductive tract. Proc Soc Exp Biol Med. 1995;209(4):325–42.
79. Toder V, Fein A, Carp H, Torchinsky A. TNF-alpha in pregnancy loss and embryo maldevelopment: a mediator of detrimental stimuli or a protector of the fetoplacental unit? J Assist Reprod Genet. 2003;20(2):73–81.
80. Argiles JM, Carbo N, Lopez-Soriano FJ. TNF and pregnancy: the paradigm of a complex interaction. Cytokine Growth Factor Rev. 1997;8(3):181–8.
81. Ntrivalas E, Kwak-Kim J, Beaman K, Mantouvalos H, Gilman-Sachs A. An in vitro coculture model to study cytokine profiles of natural killer cells during maternal immune cell-trophoblast interactions. J Soc Gynecol Investig. 2006;13(3):196–202.
82. Veenstra van Nieuwenhoven AL, Bouman A, Moes H, Heineman MJ, de Leij LF, Santema J, et al. Cytokine production in natural killer cells and lymphocytes in pregnant women compared with women in the follicular phase of the ovarian cycle. Fertil Steril. 2002;77(5):1032–7.
83. Garcia-Lloret MI, Morrish DW, Wegmann TG, Honore L, Turner AR, Guilbert LJ. Demonstration of functional cytokine-placental interactions: CSF-1 and GM-CSF stimulate human cytotrophoblast differentiation and peptide hormone secretion. Exp Cell Res. 1994;214(1):46–54.
84. Robertson SA, Roberts CT, Farr KL, Dunn AR, Seamark RF. Fertility impairment in granulocyte-macrophage colony-stimulating factor-deficient mice. Biol Reprod. 1999;60 (2):251–61.
85. Perricone R, De Carolis C, Giacomelli R, Guarino MD, De Sanctis G, Fontana L. GM-CSF and pregnancy: evidence of significantly reduced blood concentrations in unexplained recurrent abortion efficiently reverted by intravenous immunoglobulin treatment. Am J Reprod Immunol. 2003;50(3):232–7.
86. Cella M, Fuchs A, Vermi W, Facchetti F, Otero K, Lennerz JK, et al. A human natural killer cell subset provides an innate source of IL-22 for mucosal immunity. Nature. 2009;457 (7230):722–5.
87. Colonna M. Interleukin-22-producing natural killer cells and lymphoid tissue inducer-like cells in mucosal immunity. Immunity. 2009;31(1):15–23.
88. Veiga-Fernandes H, Kioussis D, Coles M. Natural killer receptors: the burden of a name. J Exp Med. 2010;207(2):269–72.
89. Yang X, Zheng SG. Interleukin-22: a likely target for treatment of autoimmune diseases. Autoimmun Rev. 2014;13(6):615–20.
90. Brosnahan MM, Miller DC, Adams M, Antczak DF. IL-22 is expressed by the invasive trophoblast of the equine (Equus caballus) chorionic girdle. J Immunol. 2012;188(9):4181–7.
91. Wang Y, Xu B, Li MQ, Li DJ, Jin LP. IL-22 secreted by decidual stromal cells and NK cells promotes the survival of human trophoblasts. Int J Clin Exp Pathol. 2013;6(9):1781–90.
92. Kamoi M, Fukui A, Kwak-Kim J, Fuchinoue K, Funamizu A, Chiba H, et al. NK22 cells in the uterine mid-secretory endometrium and peripheral blood of women with recurrent pregnancy loss and unexplained infertility. Am J Reprod Immunol. 2015;73(6):557–67.

Chapter 6
Regulation of Angiogenesis in the Human Endometrium

Hidetaka Okada, Tomoko Tsuzuki, Hiromi Murata, Atsushi Kasamatsu, Tomoo Yoshimura, and Hideharu Kanzaki

Abstract The physiological changes in the endometrium in response to hypoxia and female sex hormones are associated with profound angiogenesis. Hypoxia in the endometrial tissues is a major regulator of endometrial angiogenesis and remodeling during menstruation. Estradiol plays an important role in the reconstruction of a new vascular network and rapid vessel growth in the endometrium. Progesterone is a key factor in vascular maturation and decidualization. Further, endometrial angiogenesis is tightly controlled by a variety of angiogenic and antiangiogenic factors including vascular endothelial growth factor (VEGF), soluble VEGF receptor-1, angiopoietin, and CXCL12. Hypoxia and female sex hormones are involved in the regulation of the angiogenic factors in an independent manner in human endometrium. Analysis of the process of angiogenesis in the human endometrium will enhance our understanding of normal endometrial vascular remodeling.

Keywords Angiogenesis • Decidualization • Endometrial stromal cells • Female sex steroids • Hypoxia

6.1 Introduction

The human endometrium undergoes regular cycles of menstruation, menstrual repair, proliferation, and secretory differentiation, which are controlled by a sequential, carefully timed interplay of various environmental changes [1–3]. Endometrial blood vessels are known to grow and regress through the menstrual cycle under the control of hypoxia and female sex hormones, which act directly and indirectly via a variety of growth factors [4, 5]. The endometrium undergoes extensive growth in a cyclic manner and is regenerated nearly 450 times in a

H. Okada (✉) • T. Tsuzuki • H. Murata • A. Kasamatsu • T. Yoshimura • H. Kanzaki
Department of Obstetrics and Gynecology, Kansai Medical University, 2-5-1 shinmachi-cho, Hirakata, Osaka 573-1010, Japan
e-mail: hokada@hirakata.kmu.ac.jp

H. Kanzaki (ed.), *Uterine Endometrial Function*,
DOI 10.1007/978-4-431-55972-6_6

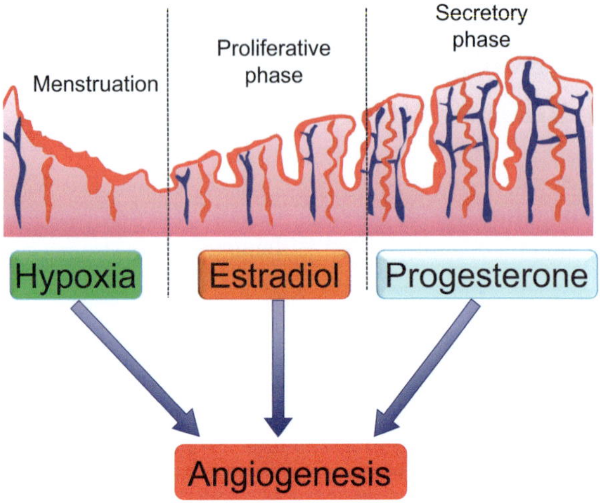

Fig. 6.1 Schematic diagram of the human menstrual cycle illustrating changes in the endometrium. The endometrium is a dynamic tissue that undergoes regular cycles in response to hypoxia, estradiol, and progesterone. The physiological changes in the endometrium during the menstrual cycle are associated with profound angiogenesis

woman's lifetime [6]. The physiological changes in the endometrium during the menstrual cycle are associated with profound angiogenesis (Fig. 6.1).

Angiogenesis is the formation of new blood vessels from existing vessels by sprouting, elongation, and intussusception from endothelial cells [7, 8]. Sprouting involves breakdown of the basement membrane, migration and proliferation of endothelial cells, tube formation, basement membrane formation, and recruitment of mural or support cells. Elongation is the lengthwise growth of vessels without formation of new vessel junctions. Intussusception involves internal division of vessels resulting in smaller vessels. These multistep processes involve activation and proliferation of endothelial cells, degradation of their basal membrane, migration through the surrounding extracellular matrix, attraction of pericellular smooth muscle cells, and maturation of vessels.

Angiogenesis occurs at all stages of the menstrual cycle [9]. It is highly regulated and critical for a number of processes such as endometrial growth and remodeling. Additionally, it plays significant roles in several gynecological disorders, including breakthrough bleeding, abnormal uterine bleeding, endometriosis, and endometrial cancer [10–12].

The first part of this chapter describes the regulation of human endometrial angiogenesis during the menstrual cycle. The second part summarizes the divergent regulation of angiogenic factors through hypoxia and female sex hormones in the human endometrium.

6.2 Regulation of Angiogenesis in the Human Endometrium

During each menstrual cycle, a new vascular system develops in the endometrium via angiogenesis and vascular remodeling to support cellular growth and differentiation. The vascular system has critical roles in homeostasis, immune defense, oxygen transport, nutrition, excretion, and fluid balance [13, 14]. The complex processes of angiogenesis are likely tightly regulated by system controls that can be turned on and off within short time.

6.2.1 *Hypoxia*

Angiogenesis is important in the cyclical repair and regeneration of the endometrium during the menstrual cycle [15]. After menstruation, the vessels of the stratum basalis are repaired following shedding of the functionalis layer of the endometrium. Hypoxia is a physiological event that occurs in the endometrial tissues during the premenstrual period. Local hypoxia is a major regulator of endometrial angiogenesis and remodeling during menstruation [16, 17].

Hypoxia-inducible factor-1 is a heterodimeric transcription factor consisting of a constitutively expressed HIF-1β subunit and a HIF-1α subunit that is regulated by the cellular O_2 concentration and plays a major role in the cellular response to hypoxia [18]. HIF-1α activation by the hypoxic microenvironment contributes to the induction or repression of the activity of genes involved in different cellular functions such as angiogenesis, cell survival, oxygen homeostasis, proliferation, glucose metabolism, and apoptosis [19, 20].

HIF-1α protein is suppressed under normoxic conditions, and its expression is rapidly induced by hypoxic conditions. In normoxia, HIF-1α is posttranslationally hydroxylated by prolyl hydroxylases in the oxygen-dependent degradation domain. Hydroxylated HIF-1α binds the von Hippel-Lindau tumor suppressor protein, which recruits an ubiquitin-ligase complex, thus targeting HIF-1α for ubiquitination and degradation by the 26S proteasome [21, 22]. Hypoxia inhibits the hydroxylation steps, which prevents the interaction of HIF-1α with the von Hippel-Lindau tumor suppressor protein and the subsequent degradation of HIF-1α. HIF-1α stabilization can be induced in the presence of chemical reagents known to mimic hypoxic conditions such as cobalt chloride ($CoCl_2$) [23, 24].

HIF-1α protein is abundant in glandular and stromal cells in the functional layer during the late secretory and menstrual phases, but not during the proliferative phase [25]. The temporal and spatial distribution of HIF-1α protein would thus seem to be related to the process of menstruation.

6.2.2 *Female Sex Steroid Hormones*

The endometrial cycle consists of two dominant phases: the proliferative phase, which follows menstruation and precedes ovulation, and the secretory phase, which occurs after ovulation. During the proliferative phase, the endometrium grows from approximately 3–9 mm in height, and the growth of new vessels must match the rapid regrowth of the functionalis [6]. During the secretory phase of the menstrual cycle, the endometrium transforms into a well-vascularized receptive tissue characterized by increased vascular permeability, edema, proliferation, invasion of leukocytes, vascular remodeling, and angiogenesis [26, 27]. Vascular changes include spiral artery remodeling, angiogenesis, and the induction of angiogenic factors. In the secretory phase, the subepithelial capillary plexus matures, and the specialized spiral arterioles grow and coil. The timing of decidual and vascular processes during the implantation period is of paramount importance for the development of a receptive endometrium suitable for implantation [28, 29].

The cycling changes occur in response to the ovarian steroid hormones estradiol (E_2) and progesterone. Hormonally controlled angiogenesis is required to support endometrial regeneration after shedding of the uterine surface in the absence of implantation and to support the proliferation of the human endometrium as well as the differentiation necessary for implantation during the menstrual cycle [5].

6.2.2.1 Estrogen

E_2 plays an important role in vascular network reconstruction and rapid vessel growth in the endometrium during each menstrual cycle [1, 5]. The proliferative phase is marked by active growth of stromal, epithelial, and vascular cells. These changes are regulated primarily by E_2 and are accompanied by changes in the expression of genes such as the progesterone receptor (PR), which permits the endometrium to respond to the progesterone produced in the secretory phase [30, 31]. Cellular responses are mediated predominantly by the estrogen receptor (ER), a member of the superfamily of ligand-inducible transcription factors [30, 32]. E_2 is clinically utilized for a variety of indications including osteoporosis and for assisted reproductive technology and hormone replacement therapy.

6.2.2.2 Progesterone

Progesterone is a key factor in the establishment and maintenance of pregnancy [33]. Cellular responses are mediated predominantly by PR [34, 35]. The endometrium is a complex multicellular tissue that undergoes dynamic remodeling in order to establish a microenvironment capable of supporting a pregnancy [36, 37]. During the establishment of pregnancy, endometrial remodeling is characterized by three key processes: influx of uterine natural killer (uNK) cells, decidualization of

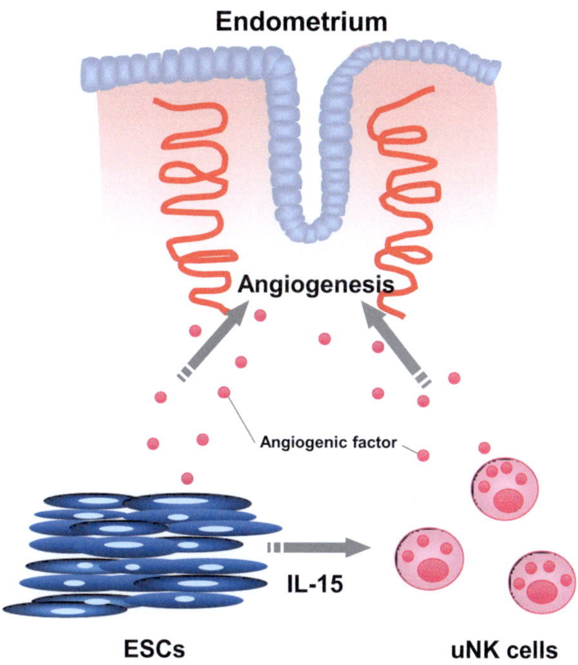

Fig. 6.2 Schematic representation showing a putative network of endometrial angiogenesis. Endometrial remodeling is characterized by three key processes: influx of uterine natural killer (uNK) cells, decidualization of endometrial stromal cells (ESCs), and remodeling of the endometrial vasculature. ESCs and uNK cells produce angiogenic factors. ESCs-derived interleukin (IL)-15 plays a role in the control of uNK cell proliferation and function

endometrial stromal cells (ESCs), and remodeling of the endometrial vasculature (Fig. 6.2).

In mouse models, uNK cells play a role in regulating uterine vascular remodeling at the implantation site [38, 39]. uNK cells contribute to pregnancy by increasing the blood flow at the fetomaternal interface and facilitating trophoblast migration [40, 41]. The main role of uNK cells is the secretion of angiogenic factors, cytokines, and growth factors [42, 43]. uNK cells and other leukocytes in the endometrium do not express PR [44]. Therefore, progesterone indirectly promotes accumulation and differentiation of uNK cells via cytokines or other soluble factors produced by ESCs, because ESCs strongly express both ER and PR. Interleukin (IL)-15 is known to play a role in the control of uNK cell proliferation and function [45–47] (Fig. 6.2). Indeed, incubation with progesterone increases synthesis and secretion of IL-15 from human ESCs in vitro [48–50].

Decidualization of the endometrium is essential for the coordinated regulation of trophoblast invasion and placenta formation. Decidualization involves dramatic morphological and functional differentiation of the ESCs and is an essential preparative event for successful establishment of pregnancy [51, 52]. Many studies have examined the regulation of and the molecular mechanisms involved in decidualization of the endometrium in vitro using human cultured ESCs expressing functional PR and ER [53, 54]. Exposure of primary ESC cultures to progesterone for 12 days triggers morphological differentiation and the expression of decidual markers such as prolactin [55]. The identification of gene expression patterns induced by specific hormones provides an insight into the molecular events underlying their diverse and tissue-specific actions [56].

6.3 Angiogenic and Antiangiogenic Factors

The development of the vascular system in the human endometrium is believed to be orchestrated by the coordinated interaction of angiogenic and antiangiogenic factors. Several angiogenic factors have been identified and are likely involved in physiological as well as pathological angiogenesis in the human endometrium (Fig. 6.3).

Vascular endothelial growth factor (VEGF) is considered the chief angiogenic growth factor in the endometrium [57–59]. Other endometrial angiogenesis promoters include angiopoietin (ANGPT), CXCL12 (stromal cell-derived factor 1; SDF-1), angiogenin, epidermal growth factor (EGF), fibroblast growth factor-2 (FGF-2), insulin-like growth factor (IGF), transforming growth factor-β1 (TGF-β1), platelet-derived growth factor (PDGF), and prokineticin-1 (PROK-1) [1, 10, 60]. Additionally, antiangiogenic factors such as endostatin, maspin, soluble VEGF receptor-1 (sVEGFR-1), and thrombospondin-1 (TSP-1) have been identified in the endometrium. Hypoxia and female sex hormones independently regulate these factors in human ESCs. Endometrial vascular remodeling is regulated by several mediators including integrins and matrix metalloproteinases (MMPs) [61, 62].

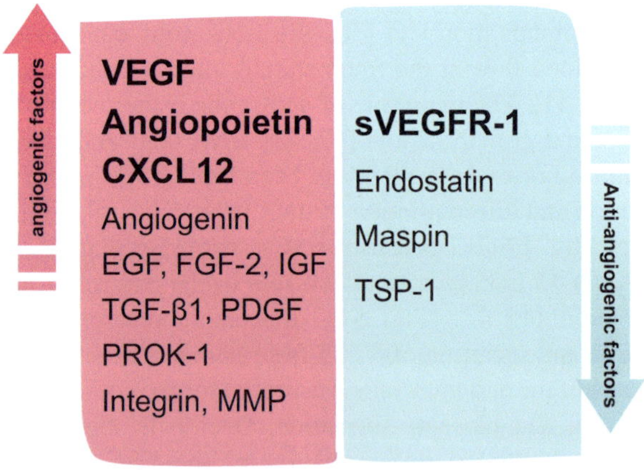

Fig. 6.3 Angiogenic and antiangiogenic factors in the human endometrium. Endometrial angiogenesis promoters include vascular endothelial growth factor (VEGF), angiopoietin (ANGPT), CXCL12, angiogenin, epidermal growth factor (EGF), fibroblast growth factor-2 (FGF-2), insulin-like growth factor (IGF), transforming growth factor-β1 (TGF-β1), platelet-derived growth factor (PDGF), prokineticin-1 (PROK-1), integrin, and matrix metalloproteinase (MMP). Antiangiogenic factors such as soluble VEGF receptor-1 (sVEGFR-1), endostatin, maspin, and thrombospondin-1 (TSP-1) have been identified in the endometrium

6.3.1 VEGF

VEGF is a key regulator of angiogenesis and vascular function. Seven VEGF subtypes (VEGF-A to VEGF-F and placental growth factor) and three VEGFR subtypes (VEGFR-1 to VEGFR-3) have been identified [63]. VEGF stimulates endothelial cell proliferation, permeability, migration, and assembly into capillary tubes [64]. Most studies on the regulation of endometrial angiogenesis have been focused on VEGF-A, usually referred to as VEGF. Alternative splicing of VEGF mRNA results in at least four polypeptide isoforms of 121, 165, 189, and 206 amino acids [65]. The most widely expressed forms, VEGF121 and VEGF165, are soluble and secreted, while VEGF189 and VEGF206 remain in the extracellular matrix.

Many studies have demonstrated expression of VEGF and all of its splice variants in the human endometrium [59, 66]. VEGF plays important roles in both physiological and pathological endometrial angiogenesis. Previous studies in mice and rabbits have shown that VEGF and its receptors probably participate in increased angiogenesis and vascular permeability required for implantation [58, 67]. Deficient expression of VEGF and its receptors in mice results in poor development of vascular network in the endometrium leading to implantation failure and abortion [68].

VEGF actions are mediated through its binding with two tyrosine kinase receptors, VEGFR-1/Flt-1 (Fms-like tyrosine kinase-1) and VEGFR-2/KDR (kinase insert domain receptor) [69, 70]. Although VEGFR-1 binds to VEGF with high affinity, most of the biological effects of VEGF are mediated by VEGFR-2. sVEGFR-1 is produced through alternative splicing and is known to act as a specific high-affinity antagonist of VEGF function by competitively binding VEGF.

The role of VEGF in the early angiogenic processes is associated with postmenstrual regeneration of the endometrium [71, 72]. VEGF is essential for the rapid burst of angiogenesis that occurs during postmenstrual repair and in the early proliferative phases and further plays a role in re-epithelialization of the endometrium in both mice and rhesus macaques [73].

Hypoxia has been shown to induce VEGF mRNA expression in human ESCs [74, 75]. In addition, increased VEGF production has been observed in ESCs after stimulation with $CoCl_2$, a chemical that induces a hypoxia-like condition by preventing proteasomal degradation of HIF-1α proteins [23]. These observations suggest that the effect of hypoxia on VEGF production in ESCs may have physiological relevance in the process of menstruation, postmenstrual endometrial repair, and angiogenesis.

VEGF is transcriptionally regulated by HIF-1α. HIF-1α binds directly to the hypoxia response elements (HREs) in the promoters of the VEGF genes [76, 77]. In fact, hypoxia has been shown to induce VEGF mRNA expression via HIF-1α activation in the human endometrium [23, 75]. Recent studies have reported that hypoxia-induced VEGF secretion is significantly attenuated by echinomycin [23]. Echinomycin is a sequence-specific DNA-binding agent that binds to the HRE site within HIF-1α target gene promoters and selectively inhibits HIF-1α

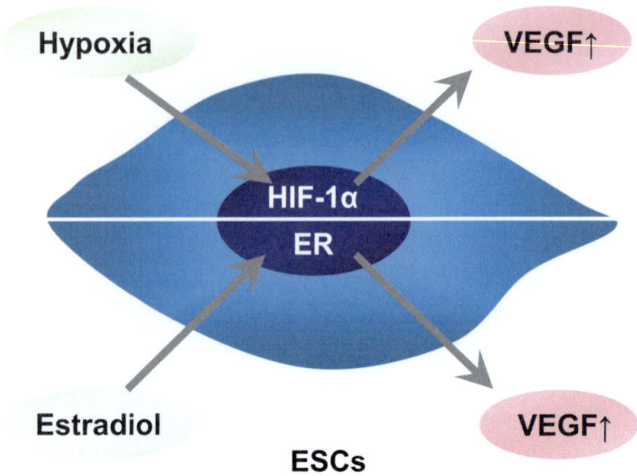

Fig. 6.4 Hypoxia and female estradiol independently regulate vascular endothelial growth factor (VEGF) in human endometrial stromal cells (ESCs). Hypoxia induces VEGF mRNA expression via hypoxia-inducible factor (HIF)-1α activation. Estradiol induces VEGF mRNA expression through the estrogen receptor (ER) in ESCs

binding activity [78]. Echinomycin can inhibit VEGF expression without changing the HIF-1α protein level and causing cell toxicity. These results indicate that hypoxia acts to increase VEGF via HIF-1α in ESCs (Fig. 6.4).

Studies on the effects of steroid hormones on endometrial vascularization have shown a marked reduction in VEGF levels in the endometrial glandular epithelial and stromal cells after oophorectomy in baboons and rhesus monkeys [79, 80]. Administration of E_2 restored VEGF expression and microvascular permeability, which are the early events of angiogenesis [5]. E_2 induces VEGF production through the ER in ESCs [81]. In cultured ESCs, E_2 induces VEGF mRNA and protein production but attenuates sVEGFR-1 mRNA and protein production [82]. The latter results are in agreement with a recent study describing decreased sVEGFR-1 expression in an ER-positive cell line following treatment with E_2 [83, 84]. The E_2-regulated decrease in sVEGFR-1 expression was accompanied by a significant increase in angiogenesis. Evidence suggests that sVEGFR-1 can affect endometrial maturation by directly affecting angiogenesis. sVEGFR-1 can substantially modify the responses of the endometrium to steroids through a direct effect on the endometrium, independent of the corpus luteum, in ovariectomized mice [85]. As sVEGFR-1 is a VEGF antagonist, the actual angiogenic potential of the VEGF system depends on the balance between VEGF and sVEGFR-1. Therefore, the increase in the VEGF/sVEGFR-1 ratio following treatment with E_2 appears to be a sustained and ongoing process promoting growth and development of the endometrium during the advancing stages of the menstrual cycle at the local level.

Increased VEGF production by ESCs upon E_2 treatment is brought about, at least in part, at the pretranslational level as judged by an increase in VEGF mRNA. The increase in VEGF mRNA expression is likely mediated by the ER, because E_2 stimulates VEGF gene transcription through a functional variant estrogen response element on the VEGF promoter [86] (Fig. 6.4).

Progesterone and medroxyprogesterone acetate (MPA) inhibit the E_2-induced increase in VEGF in ESCs [87]. Co-treatment with the PR antagonist RU-486 reverses the inhibition of E_2-stimulated VEGF production, suggesting a pathway by which progestins may reduce VEGF production through the PR. Progestins are known to initiate downregulation of the ER in the human endometrium in vivo as well as in vitro; therefore, the inhibition of VEGF may be caused by decreased ER levels [88–90]. Alternatively, ligand-occupied ER and PR bind directly to DNA at the steroid response element and recruit co-regulators that activate or repress transcription via interaction with the general transcription apparatus [91]. PR has been shown to inhibit ER-dependent gene activation presumably through competition for common limiting coactivators. Together, these findings suggest the complexity of the PR and ER signaling pathways.

6.3.2 ANGPT

ANGPTs comprise a second key group of promoters of angiogenesis and vessel remodeling in the endometrium and interact with VEGF. ANGPTs appear to play major roles in the regulation of blood vessel growth, maturation, and regression [92]. ANGPTs include four subtypes (ANGPT1, ANGPT2, ANGPT3, and ANGPT4), of which the best characterized are ANGPT1 and ANGPT2 [93, 94]. ANGPT2 is selectively expressed in the ovary, uterus, and placenta, which are known as tissues that undergo physiological angiogenesis, whereas ANGPT1 is widely expressed [95, 96]. In the human endometrium, ANGPT1 is mainly expressed in the stroma surrounding the blood vessels in the secretory phase, whereas ANGPT2 expression is mainly localized to the glandular epithelium and endothelium [97].

Both ANGPT1 and ANGPT2 act on vascular endothelial cells via a single receptor tyrosine kinase (TIE-2) with immunoglobulin-like and EGF-like domains, with similar affinity [96, 98]. Although ANGPT1 and ANGPT2 share a similar protein structure, their biological activities differ significantly. ANGPT1 acts as a paracrine agonist to TIE-2, leads to receptor dimerization, and induces its phosphorylation on several cytoplasmic residues to activate downstream signaling pathways [99]. ANGPT1 increases the association of endothelial cells with pericytes and vascular smooth muscle cells to stimulate the maturation of the vascular network. Conversely, ANGPT2 is a natural antagonist of ANGPT1 that promotes loosening of the supporting cell matrix and destabilization of existing vessels and initiates neovascularization in the presence of VEGF [100, 101]. In the absence of VEGF, ANGPT2 blocks the recruitment of periendothelial support cells, resulting in blood vessel destabilization and regression.

The balance between ANGPT1 and ANGPT2 expression is important for the formation, development, and stabilization of blood vessels. In addition, several reports have favored the ANGPT2/ANGPT1 ratio (an index of vasculature instability) as a more optimal reflection of ANGPT balance [102, 103]. An increase in the ANGPT2/ANGPT1 ratio is associated with blood vessel destabilization, which is a prerequisite for new blood vessel formation [104].

Fig. 6.5 The summary of the divergent regulation of angiogenic factors through hypoxia, estradiol (E_2), and progesterone (P). Hypoxia and female sex hormones are involved in the regulation of vascular endothelial growth factor (VEGF), angiopoietin (ANGPT), CXCL12 in an independent manner in human endometrial stromal cells

	Hypoxia	E_2	E_2+P
VEGF	↑↑	↑↑	→
ANGPT1	↓	↓	→
ANGPT2	→	→	↓↓
ANGPT2/ ANGPT1	↑	↑	↓↓
CXCL12	↓	↑↑	→

In human ESCs, hypoxia reduces ANGPT1 production and maintains ANGPT2, resulting in an increase of the ANGPT2/ANGPT1 ratio [105]. Hypoxia simultaneously induces VEGF production. The increase in the ANGPT2/ANGPT1 ratio in the presence of VEGF is associated with new blood vessel formation. Thus, the changes in ESCs under hypoxic conditions may contribute to the formation of additional blood vessels in the endometrium.

While E_2 induces VEGF production, it simultaneously suppresses ANGPT1 production, resulting in an increase of the ANGPT2/ANGPT1 ratio in human ESCs (Fig. 6.5). Similarly, treatment with E_2 causes a significant increase in VEGF expression and a decrease in ANGPT1 expression in placental villous trophoblasts [106]. The increase in the ANGPT2/ANGPT1 ratio and VEGF production in ESCs following E_2 treatment may create an environment for the development, reconstruction, and remodeling of the endometrial blood vessel network. Recent reports have indicated that E_2 reduces ANGPT1 mRNA and protein expression in ER-positive, but not in ER-negative breast cancer cells [107]. However, in baboons, ANGPT1 transcription has been shown to increase in the endometrium after E_2 treatment [108].

Vascular maturation occurs during the secretory phase of the menstrual cycle and is regulated by progesterone. Progestins have been shown to reduce ANGPT2 production and sustain the levels of ANGPT1, resulting in a decrease of the ANGPT2/ANGPT1 ratio [105]. These results are in agreement with a recent study showing that progesterone can inhibit ANGPT2 expression in ovariectomized mouse uterus [109]. The dominance of ANGPT1 and lower ANGPT2/ANGPT1 ratio following treatment of ESCs with MPA appear to favor the maturation and stabilization of newly developed vessels in the endometrium.

Previous studies have suggested that hypoxia and E_2 can cooperate to regulate gene expression in human breast cancer and endometrial cancer cells [110]. However, E_2 did not affect the regulation of VEGF, ANGPT1, and ANGPT2 secretion under hypoxic conditions in ESCs. We found that the induction of HIF-1α protein in hypoxia did not change with the addition of E_2 in human ESCs [105]. These findings are interesting because E_2 has been demonstrated to induce an increase in HIF-1α levels in human endometrial cancer cells and in rat uterus [111, 112]. HIF-1α is expressed with increasing intensity in the menstrual phase in the human endometrium, but not during the proliferative phase regulated by E_2. These findings suggest that hypoxia and sex hormones independently regulate the angiogenic factors in ESCs (Fig. 6.3).

6.3.3 *CXCL12*

CXCL12 (SDF-1) is a member of the CXC chemokine family that was initially cloned from mouse bone marrow and has been characterized as a pre-B-cell growth-stimulating factor [113]. The effects of CXCL12 are mediated by its interaction with CXC chemokine receptor-4 (CXCR-4), which is the only physiological receptor for CXCL12 [114].

CXCL12 plays an important role in angiogenesis and is a potent chemoattractant for leukocytes, endothelial cells, and hematopoietic progenitor cells [115]. CXCL12 and VEGF act synergistically to promote the functions of vascular endothelial cells such as cell survival and migration and changes in gene expression. VEGF enhances the expression of CXCR4, and conversely, CXCL12 enhances the production of VEGF in a positive feedback loop in human umbilical vein endothelial cells [116, 117].

It has been shown to be upregulated in multiple damaged tissues as part of the injury response and is thought to channel stem and progenitor cells to promote repair. Recent studies have indicated the importance of CXCL12 in human endometrial function. CXCL12 may play a crucial role in endometrial proliferation, leukocyte recruitment, and embryo implantation [118, 119].

Hypoxia attenuates the expression and production of CXCL12 in ESCs in a time-dependent manner [23] (Fig. 6.5). Similar dose-dependent changes have been observed in ESCs treated with the hypoxia-mimicking agent $CoCl_2$. These findings are surprising because hypoxia has been shown to induce the production of CXCL12 in cancer cells and endothelial cells [120, 121]. Echinomycin has no effect on this suppression. Echinomycin specifically inhibits HIF-1α, but not activator protein 1 (AP-1)- and nuclear factor (NF)-κB-dependent DNA-binding activity [78]. Hypoxia induces the activation of the transcription factors HIF-1α, AP-1, and NF-κB [122, 123]. It has been speculated that the hypoxia-induced inhibition of CXCL12 production is caused by the action of AP-1 or NF-κB.

E_2 enhances CXCL12 production in ESCs in a dose- and time-dependent manner [119] (Fig. 6.5). In addition, E_2-stimulated CXCL12 production is blocked by co-treatment of cells with the specific ER antagonist ICI 182,780, suggesting a pathway by which E_2 may induce CXCL12 production through the ER. Indeed, the CXCL12 promoter contains an estrogen response element half-site [124]. Although

E_2 significantly increases the levels of CXCL12 production in human ESCs, CXCL12 production is undetectable in Ishikawa cells. Ishikawa cells are a well-differentiated human endometrial epithelial cell line and are widely considered a good model for the study of normal endometrial function. CXCL12 induces proliferation of Ishikawa cells, which can be inhibited by AMD3100, an antagonist of CXCR4 [119]. E_2-induced CXCL12 mediates proliferation of Ishikawa cells via CXCR4 interaction in a paracrine manner. Therefore, CXCL12 secreted by ESCs is regarded a candidate factor involved in the cross talk of the epithelial-stromal interaction in the human endometrium and may be actively involved in its proliferation.

Progesterone and MPA antagonize E_2-stimulated CXCL12 production in a time- and dose-dependent manner, and this effect is reversed by RU-486 [87]. Steroid hormones rather than hypoxia may be the main regulator of CXCL12 secretion in ESCs, indicating that CXCL12 plays a more important role in endometrial proliferation than in repair during the perimenstrual period.

6.3.4 Other Angiogenic Factors

Angiogenin is a heparin-binding 14.1-kDa single-chain polypeptide with high angiogenic activity [125]. Angiogenin is expressed in the human endometrium, with enhanced expression during the secretory phase and in decidual tissues, raising the possibility that angiogenin may play a role in establishing pregnancy [126]. Angiogenin may contribute to preventing vessel breakdown in the initial stages of spiral artery remodeling during pregnancy.

EGF, FGF2, and IGF have been recognized as important mitogens for angiogenesis, which promote endothelial cell proliferation and differentiation [127, 128]. Expression of these factors in human endometrium has been verified. TGF-β1 has been proposed to play roles in stabilizing newly formed vessels as well as in directing vascular smooth muscle cell differentiation [129]. The PDGF family has four members (PDGF-A, PDGF-B, PDGF-C, and PDGF-D), which form homo- and heterodimers. The homodimer PDGF-BB appears to play a key role in arteriogenesis [130]. Mice deficient in PDGF-BB have diminished numbers of pericytes and markedly dilated blood vessels, leading to edema and embryonic lethality [131]. Those expression profiles have been previously investigated in human endometrium [127, 132, 133].

PROK1 is an angiogenic factor implicated in the vascular function of peri-implantation endometrium and early pregnancy [134]. PROK1 displays differential expression throughout the menstrual cycle with elevation in the secretory phase, and expression is regulated by steroid hormones [135]. PROK1 localizes to the glandular and luminal epithelium, endothelial, and stromal cells of the endometrium [136].

Endostatin is the C-terminal cleavage product of collagen XVIII, an extracellular matrix protein associated with the basement membrane [137, 138]. It is an antiangiogenic factor that most likely works through several modes of action including direct interaction with the basement membrane, inhibition of MMPs, and interaction with endothelial cell receptors [139, 140]. Endostatin is expressed in the decidua and in cultured endometrial stromal cells [141, 142].

Maspin has an important role in the inhibition of angiogenesis, tumor cell motility, adhesion, invasion, and migration [143, 144]. In mice, maspin expression gradually increases during the early days of pregnancy and reaches a maximum at the time of implantation [145]. These findings suggest that maspin plays an important role in the regulation of early embryonic implantation in the endometrium.

TSP-1 is a potent regulator of angiogenesis that concurrently inhibits endothelial cell migration and release of VEGF from the extracellular matrix [146, 147]. TSP-1 expression is markedly higher in secretory-phase endometrium than in proliferative-phase endometrium [148]. E_2 inhibits both mRNA and protein expression of TSP-1 in ESCs [149].

6.4 Conclusions

Local levels of autocrine and paracrine molecules vary during the menstrual cycle and have been suggested to play various roles in endometrial function. Hypoxia and female sex hormones are involved in the regulation of angiogenic regulators in an independent manner in human ESCs (Fig. 6.5). The changes in these angiogenic factors may be related to the environment of the human endometrium associated with the menstrual cycle and may be indicative of ingenious hypoxic or hormonal control. Further studies on the process of angiogenesis in the human endometrium will ultimately enhance our understanding of normal and pathological endometrial vascular remodeling.

References

1. Okada H, Tsuzuki T, Shindoh H, Nishigaki A, Yasuda K, Kanzaki H. Regulation of decidualization and angiogenesis in the human endometrium: mini review. J Obstet Gynaecol Res. 2014;40:1180–7.
2. Lessey BA, Young SL. Homeostasis imbalance in the endometrium of women with implantation defects: the role of estrogen and progesterone. Semin Reprod Med. 2014;32:365–75.
3. Henriet P, Gaide Chevronnay HP, Marbaix E. The endocrine and paracrine control of menstruation. Mol Cell Endocrinol. 2012;358:197–207.
4. Plaisier M. Decidualisation and angiogenesis. Best Pract Res Clin Obstet Gynaecol. 2011;25:259–71.
5. Albrecht ED, Pepe GJ. Steroid hormone regulation of angiogenesis in the primate endometrium. Front Biosci. 2003;8:d416–29.

6. Petracco RG, Kong A, Grechukhina O, Krikun G, Taylor HS. Global gene expression profiling of proliferative phase endometrium reveals distinct functional subdivisions. Reprod Sci. 2012;19:1138–45.
7. Carmeliet P. Angiogenesis in health and disease. Nat Med. 2003;9:653–60.
8. Girling JE, Rogers PA. Recent advances in endometrial angiogenesis research. Angiogenesis. 2005;8:89–99.
9. Smith SK. Angiogenesis and implantation. Hum Reprod. 2000;15 Suppl 6:59–66.
10. Hey-Cunningham AJ, Peters KM, Zevallos HB, Berbic M, Markham R, Fraser IS. Angiogenesis, lymphangiogenesis and neurogenesis in endometriosis. Front Biosci (Elite Ed). 2013;5:1033–56.
11. Pittatore G, Moggio A, Benedetto C, Bussolati B, Revelli A. Endometrial adult/progenitor stem cells: pathogenetic theory and new antiangiogenic approach for endometriosis therapy. Reprod Sci. 2014;21:296–304.
12. Bradford LS, Rauh-Hain JA, Schorge J, Birrer MJ, Dizon DS. Advances in the management of recurrent endometrial cancer. Am J Clin Oncol. 2015;38:206–12.
13. Virdis A, Dell'Agnello U, Taddei S. Impact of inflammation on vascular disease in hypertension. Maturitas. 2014;78:179–83.
14. Schulte-Merker S, Sabine A, Petrova TV. Lymphatic vascular morphogenesis in development, physiology, and disease. J Cell Biol. 2011;193:607–18.
15. Jabbour HN, Kelly RW, Fraser HM, Critchley HO. Endocrine regulation of menstruation. Endocr Rev. 2006;27:17–46.
16. Aberdeen GW, Wiegand SJ, Bonagura Jr TW, Pepe GJ, Albrecht ED. Vascular endothelial growth factor mediates the estrogen-induced breakdown of tight junctions between and increase in proliferation of microvessel endothelial cells in the baboon endometrium. Endocrinology. 2008;149:6076–83.
17. Salamonsen LA. Tissue injury and repair in the female human reproductive tract. Reproduction. 2003;125:301–11.
18. Majmundar AJ, Wong WJ, Simon MC. Hypoxia-inducible factors and the response to hypoxic stress. Mol Cell. 2010;40:294–309.
19. Semenza GL. Oxygen sensing, homeostasis, and disease. N Engl J Med. 2011;365:537–47.
20. Szablowska-Gadomska I, Zayat V, Buzanska L. Influence of low oxygen tensions on expression of pluripotency genes in stem cells. Acta Neurobiol Exp (Wars). 2011;71:86–93.
21. Ivan M, Kondo K, Yang H, Kim W, Valiando J, Ohh M, Salic A, Asara JM, Lane WS, Kaelin Jr WG. Hifalpha targeted for vhl-mediated destruction by proline hydroxylation: implications for o2 sensing. Science. 2001;292:464–8.
22. Jaakkola P, Mole DR, Tian YM, Wilson MI, Gielbert J, Gaskell SJ, von Kriegsheim A, Hebestreit HF, Mukherji M, Schofield CJ, Maxwell PH, Pugh CW, Ratcliffe PJ. Targeting of hif-alpha to the von hippel-lindau ubiquitylation complex by o2-regulated prolyl hydroxylation. Science. 2001;292:468–72.
23. Tsuzuki T, Okada H, Cho H, Tsuji S, Nishigaki A, Yasuda K, Kanzaki H. Hypoxic stress simultaneously stimulates vascular endothelial growth factor via hypoxia-inducible factor-1alpha and inhibits stromal cell-derived factor-1 in human endometrial stromal cells. Hum Reprod. 2012;27:523–30.
24. Ardyanto TD, Osaki M, Tokuyasu N, Nagahama Y, Ito H. Cocl2-induced hif-1alpha expression correlates with proliferation and apoptosis in mkn-1 cells: a possible role for the pi3k/akt pathway. Int J Oncol. 2006;29:549–55.
25. Critchley HO, Osei J, Henderson TA, Boswell L, Sales KJ, Jabbour HN, Hirani N. Hypoxia-inducible factor-1alpha expression in human endometrium and its regulation by prostaglandin e-series prostanoid receptor 2 (ep2). Endocrinology. 2006;147:744–53.
26. Lobo SC, Huang ST, Germeyer A, Dosiou C, Vo KC, Tulac S, Nayak NR, Giudice LC. The immune environment in human endometrium during the window of implantation. Am J Reprod Immunol. 2004;52:244–51.

27. Okada H, Nie G, Salamonsen LA. Requirement for proprotein convertase 5/6 during decidualization of human endometrial stromal cells in vitro. J Clin Endocrinol Metab. 2005;90:1028–34.
28. Dimitriadis E, Nie G, Hannan NJ, Paiva P, Salamonsen LA. Local regulation of implantation at the human fetal-maternal interface. Int J Dev Biol. 2010;54:313–22.
29. Singh M, Chaudhry P, Asselin E. Bridging endometrial receptivity and implantation: network of hormones, cytokines, and growth factors. J Endocrinol. 2011;210:5–14.
30. Hamilton KJ, Arao Y, Korach KS. Estrogen hormone physiology: reproductive findings from estrogen receptor mutant mice. Reprod Biol. 2014;14:3–8.
31. Koos RD. Minireview: putting physiology back into estrogens' mechanism of action. Endocrinology. 2011;152:4481–8.
32. Hara Y, Waters EM, McEwen BS, Morrison JH. Estrogen effects on cognitive and synaptic health over the lifecourse. Physiol Rev. 2015;95:785–807.
33. Maruyama T, Yoshimura Y. Molecular and cellular mechanisms for differentiation and regeneration of the uterine endometrium. Endocr J. 2008;55:795–810.
34. Conneely OM, Mulac-Jericevic B, DeMayo F, Lydon JP, O'Malley BW. Reproductive functions of progesterone receptors. Recent Prog Horm Res. 2002;57:339–55.
35. Wetendorf M, DeMayo FJ. The progesterone receptor regulates implantation, decidualization, and glandular development via a complex paracrine signaling network. Mol Cell Endocrinol. 2012;357:108–18.
36. Lee SK, Kim CJ, Kim DJ, Kang JH. Immune cells in the female reproductive tract. Immune Netw. 2015;15:16–26.
37. Wang H, Dey SK. Roadmap to embryo implantation: clues from mouse models. Nat Rev Genet. 2006;7:185–99.
38. Croy BA, Chen Z, Hofmann AP, Lord EM, Sedlacek AL, Gerber SA. Imaging of vascular development in early mouse decidua and its association with leukocytes and trophoblasts. Biol Reprod. 2012;87:125.
39. Hofmann AP, Gerber SA, Croy BA. Uterine natural killer cells pace early development of mouse decidua basalis. Mol Hum Reprod. 2014;20:66–76.
40. Lima PD, Zhang J, Dunk C, Lye SJ, Croy BA. Leukocyte driven-decidual angiogenesis in early pregnancy. Cell Mol Immunol. 2014;11:522–37.
41. Wallace AE, Fraser R, Gurung S, Goulwara SS, Whitley GS, Johnstone AP, Cartwright JE. Increased angiogenic factor secretion by decidual natural killer cells from pregnancies with high uterine artery resistance alters trophoblast function. Hum Reprod. 2014;29:652–60.
42. Hanna J, Goldman-Wohl D, Hamani Y, Avraham I, Greenfield C, Natanson-Yaron S, Prus D, Cohen-Daniel L, Arnon TI, Manaster I, Gazit R, Yutkin V, Benharroch D, Porgador A, Keshet E, Yagel S, Mandelboim O. Decidual nk cells regulate key developmental processes at the human fetal-maternal interface. Nat Med. 2006;12:1065–74.
43. Wallace AE, Fraser R, Cartwright JE. Extravillous trophoblast and decidual natural killer cells: a remodelling partnership. Hum Reprod Update. 2012;18:458–71.
44. Henderson TA, Saunders PT, Moffett-King A, Groome NP, Critchley HO. Steroid receptor expression in uterine natural killer cells. J Clin Endocrinol Metab. 2003;88:440–9.
45. Kitaya K, Yamaguchi T, Honjo H. Central role of interleukin-15 in postovulatory recruitment of peripheral blood cd16(-) natural killer cells into human endometrium. J Clin Endocrinol Metab. 2005;90:2932–40.
46. Verbist KC, Klonowski KD. Functions of il-15 in anti-viral immunity: multiplicity and variety. Cytokine. 2012;59:467–78.
47. Santoni A, Carlino C, Gismondi A. Uterine nk cell development, migration and function. Reprod Biomed Online. 2008;16:202–10.
48. Okada H, Nakajima T, Sanezumi M, Ikuta A, Yasuda K, Kanzaki H. Progesterone enhances interleukin-15 production in human endometrial stromal cells in vitro. J Clin Endocrinol Metab. 2000;85:4765–70.

49. Okada S, Okada H, Sanezumi M, Nakajima T, Yasuda K, Kanzaki H. Expression of interleukin-15 in human endometrium and decidua. Mol Hum Reprod. 2000;6:75–80.
50. Okada H, Nakajima T, Yasuda K, Kanzaki H. Interleukin-1 inhibits interleukin-15 production by progesterone during in vitro decidualization in human. J Reprod Immunol. 2004;61:3–12.
51. Ramathal CY, Bagchi IC, Taylor RN, Bagchi MK. Endometrial decidualization: of mice and men. Semin Reprod Med. 2010;28:17–26.
52. Cho H, Okada H, Tsuzuki T, Nishigaki A, Yasuda K, Kanzaki H. Progestin-induced heart and neural crest derivatives expressed transcript 2 is associated with fibulin-1 expression in human endometrial stromal cells. Fertil Steril. 2013;99:248–55.
53. Okada H, Nakajima T, Yoshimura T, Yasuda K, Kanzaki H. The inhibitory effect of dienogest, a synthetic steroid, on the growth of human endometrial stromal cells in vitro. Mol Hum Reprod. 2001;7:341–7.
54. Shindoh H, Okada H, Tsuzuki T, Nishigaki A, Kanzaki H. Requirement of heart and neural crest derivatives-expressed transcript 2 during decidualization of human endometrial stromal cells in vitro. Fertil Steril. 2014;2014(101):1781–90. e1781–1785.
55. Okada H, Sanezumi M, Nakajima T, Okada S, Yasuda K, Kanzaki H. Rapid down-regulation of cd63 transcription by progesterone in human endometrial stromal cells. Mol Hum Reprod. 1999;5:554–8.
56. Okada H, Nakajima T, Yoshimura T, Yasuda K, Kanzaki H. Microarray analysis of genes controlled by progesterone in human endometrial stromal cells in vitro. Gynecol Endocrinol. 2003;17:271–80.
57. Binder NK, Evans J, Gardner DK, Salamonsen LA, Hannan NJ. Endometrial signals improve embryo outcome: functional role of vascular endothelial growth factor isoforms on embryo development and implantation in mice. Hum Reprod. 2014;29:2278–86.
58. Chakraborty I, Das SK, Dey SK. Differential expression of vascular endothelial growth factor and its receptor mrnas in the mouse uterus around the time of implantation. J Endocrinol. 1995;147:339–52.
59. Charnock-Jones DS, Sharkey AM, Rajput-Williams J, Burch D, Schofield JP, Fountain SA, Boocock CA, Smith SK. Identification and localization of alternately spliced mrnas for vascular endothelial growth factor in human uterus and estrogen regulation in endometrial carcinoma cell lines. Biol Reprod. 1993;48:1120–8.
60. Weston G, Rogers PA. Endometrial angiogenesis. Baillieres Best Pract Res Clin Obstet Gynaecol. 2000;14:919–36.
61. Gargett CE, Rogers PA. Human endometrial angiogenesis. Reproduction. 2001;121:181–6.
62. Smith SK. Regulation of angiogenesis in the endometrium. Trends Endocrinol Metab. 2001;12:147–51.
63. Shibuya M. Differential roles of vascular endothelial growth factor receptor-1 and receptor-2 in angiogenesis. J Biochem Mol Biol. 2006;39:469–78.
64. Nishigaki A, Okada H, Okamoto R, Shimoi K, Miyashiro H, Yasuda K, Kanzaki H. The concentration of human follicular fluid stromal cell-derived factor-1 is correlated with luteinization in follicles. Gynecol Endocrinol. 2013;29:230–4.
65. Ferrara N. Vascular endothelial growth factor: basic science and clinical progress. Endocr Rev. 2004;25:581–611.
66. Shifren JL, Tseng JF, Zaloudek CJ, Ryan IP, Meng YG, Ferrara N, Jaffe RB, Taylor RN. Ovarian steroid regulation of vascular endothelial growth factor in the human endometrium: implications for angiogenesis during the menstrual cycle and in the pathogenesis of endometriosis. J Clin Endocrinol Metab. 1996;81:3112–8.
67. Das SK, Chakraborty I, Wang J, Dey SK, Hoffman LH. Expression of vascular endothelial growth factor (vegf) and vegf-receptor messenger ribonucleic acids in the peri-implantation rabbit uterus. Biol Reprod. 1997;56:1390–9.
68. Evans PW, Wheeler T, Anthony FW, Osmond C. A longitudinal study of maternal serum vascular endothelial growth factor in early pregnancy. Hum Reprod. 1998;13:1057–62.
69. Shibuya M. Vegf-vegfr signals in health and disease. Biomol Ther (Seoul). 2014;22:1–9.

70. Wittko-Schneider IM, Schneider FT, Plate KH. Brain homeostasis: vegf receptor 1 and 2-two unequal brothers in mind. Cell Mol Life Sci. 2013;70:1705–25.
71. Lockwood CJ. Mechanisms of normal and abnormal endometrial bleeding. Menopause. 2011;18:408–11.
72. Becker CM, D'Amato RJ. Angiogenesis and antiangiogenic therapy in endometriosis. Microvasc Res. 2007;74:121–30.
73. Fan X, Krieg S, Kuo CJ, Wiegand SJ, Rabinovitch M, Druzin ML, Brenner RM, Giudice LC, Nayak NR. Vegf blockade inhibits angiogenesis and reepithelialization of endometrium. FASEB J. 2008;22:3571–80.
74. Sharkey AM, Day K, McPherson A, Malik S, Licence D, Smith SK, Charnock-Jones DS. Vascular endothelial growth factor expression in human endometrium is regulated by hypoxia. J Clin Endocrinol Metab. 2000;85:402–9.
75. Yoshie M, Miyajima E, Kyo S, Tamura K. Stathmin, a microtubule regulatory protein, is associated with hypoxia-inducible factor-1alpha levels in human endometrial and endothelial cells. Endocrinology. 2009;150:2413–8.
76. Breen E, Tang K, Olfert M, Knapp A, Wagner P. Skeletal muscle capillarity during hypoxia: vegf and its activation. High Alt Med Biol. 2008;9:158–66.
77. Kimura H, Weisz A, Kurashima Y, Hashimoto K, Ogura T, D'Acquisto F, Addeo R, Makuuchi M, Esumi H. Hypoxia response element of the human vascular endothelial growth factor gene mediates transcriptional regulation by nitric oxide: control of hypoxia-inducible factor-1 activity by nitric oxide. Blood. 2000;95:189–97.
78. Kong D, Park EJ, Stephen AG, Calvani M, Cardellina JH, Monks A, Fisher RJ, Shoemaker RH, Melillo G. Echinomycin, a small-molecule inhibitor of hypoxia-inducible factor-1 DNA-binding activity. Cancer Res. 2005;65:9047–55.
79. Niklaus AL, Aberdeen GW, Babischkin JS, Pepe GJ, Albrecht ED. Effect of estrogen on vascular endothelial growth/permeability factor expression by glandular epithelial and stromal cells in the baboon endometrium. Biol Reprod. 2003;68:1997–2004.
80. Nayak NR, Brenner RM. Vascular proliferation and vascular endothelial growth factor expression in the rhesus macaque endometrium. J Clin Endocrinol Metab. 2002;87:1845–55.
81. Herve MA, Meduri G, Petit FG, Domet TS, Lazennec G, Mourah S, Perrot-Applanat M. Regulation of the vascular endothelial growth factor (vegf) receptor flk-1/kdr by estradiol through vegf in uterus. J Endocrinol. 2006;188:91–9.
82. Okada H, Tsutsumi A, Imai M, Nakajima T, Yasuda K, Kanzaki H. Estrogen and selective estrogen receptor modulators regulate vascular endothelial growth factor and soluble vascular endothelial growth factor receptor 1 in human endometrial stromal cells. Fertil Steril. 2010;93:2680–6.
83. Elkin M, Orgel A, Kleinman HK. An angiogenic switch in breast cancer involves estrogen and soluble vascular endothelial growth factor receptor 1. J Natl Cancer Inst. 2004;96:875–8.
84. Garvin S, Nilsson UW, Dabrosin C. Effects of oestradiol and tamoxifen on vegf, soluble vegfr-1, and vegfr-2 in breast cancer and endothelial cells. Br J Cancer. 2005;93:1005–10.
85. Sharkey AM, Catalano R, Evans A, Charnock-Jones DS, Smith SK. Novel antiangiogenic agents for use in contraception. Contraception. 2005;71:263–71.
86. Mueller MD, Vigne JL, Minchenko A, Lebovic DI, Leitman DC, Taylor RN. Regulation of vascular endothelial growth factor (vegf) gene transcription by estrogen receptors alpha and beta. Proc Natl Acad Sci U S A. 2000;97:10972–7.
87. Okada H, Okamoto R, Tsuzuki T, Tsuji S, Yasuda K, Kanzaki H. Progestins inhibit estradiol-induced vascular endothelial growth factor and stromal cell-derived factor 1 in human endometrial stromal cells. Fertil Steril. 2011;96:786–91.
88. Smith OP, Critchley HO. Progestogen only contraception and endometrial break through bleeding. Angiogenesis. 2005;8:117–26.
89. Vereide AB, Kaino T, Sager G, Arnes M, Orbo A. Effect of levonorgestrel iud and oral medroxyprogesterone acetate on glandular and stromal progesterone receptors (pra and prb),

and estrogen receptors (er-alpha and er-beta) in human endometrial hyperplasia. Gynecol Oncol. 2006;101:214–23.
90. Hanifi-Moghaddam P, Sijmons B, Ott MC, van Ijcken WF, Nowzari D, Kuhne EC, van der Spek P, Kloosterboer HJ, Burger CW, Blok LJ. The hormone replacement therapy drug tibolone acts very similar to medroxyprogesterone acetate in an estrogen-and progesterone-responsive endometrial cancer cell line. J Mol Endocrinol. 2006;37:405–13.
91. Conneely OM, Lydon JP. Progesterone receptors in reproduction: functional impact of the a and b isoforms. Steroids. 2000;65:571–7.
92. Nishigaki A, Okada H, Tsuzuki T, Cho H, Yasuda K, Kanzaki H. Angiopoietin 1 and angiopoietin 2 in follicular fluid of women undergoing a long protocol. Fertil Steril. 2011;96:1378–83.
93. Thurston G. Role of angiopoietins and tie receptor tyrosine kinases in angiogenesis and lymphangiogenesis. Cell Tissue Res. 2003;314:61–8.
94. Thomas M, Augustin HG. The role of the angiopoietins in vascular morphogenesis. Angiogenesis. 2009;12:125–37.
95. Gale NW, Yancopoulos GD. Growth factors acting via endothelial cell-specific receptor tyrosine kinases: vegfs, angiopoietins, and ephrins in vascular development. Genes Dev. 1999;13:1055–66.
96. Davis S, Aldrich TH, Jones PF, Acheson A, Compton DL, Jain V, Ryan TE, Bruno J, Radziejewski C, Maisonpierre PC, Yancopoulos GD. Isolation of angiopoietin-1, a ligand for the tie2 receptor, by secretion-trap expression cloning. Cell. 1996;87:1161–9.
97. Hewett P, Nijjar S, Shams M, Morgan S, Gupta J, Ahmed A. Down-regulation of angiopoietin-1 expression in menorrhagia. Am J Pathol. 2002;160:773–80.
98. Maisonpierre PC, Suri C, Jones PF, Bartunkova S, Wiegand SJ, Radziejewski C, Compton D, McClain J, Aldrich TH, Papadopoulos N, Daly TJ, Davis S, Sato TN, Yancopoulos GD. - Angiopoietin-2, a natural antagonist for tie2 that disrupts in vivo angiogenesis. Science. 1997;277:55–60.
99. Khan AA, Sandhya VK, Singh P, Parthasarathy D, Kumar A, Advani J, Gattu R, Ranjit DV, Vaidyanathan R, Mathur PP, Prasad TS, Mac Gabhann F, Pandey A, Raju R, Gowda H. Signaling network map of endothelial tek tyrosine kinase. J Signal Transduct. 2014;2014:173026.
100. Thurston G, Rudge JS, Ioffe E, Zhou H, Ross L, Croll SD, Glazer N, Holash J, McDonald DM, Yancopoulos GD. Angiopoietin-1 protects the adult vasculature against plasma leakage. Nat Med. 2000;6:460–3.
101. Hashizume H, Falcon BL, Kuroda T, Baluk P, Coxon A, Yu D, Bready JV, Oliner JD, McDonald DM. Complementary actions of inhibitors of angiopoietin-2 and vegf on tumor angiogenesis and growth. Cancer Res. 2010;70:2213–23.
102. Conde-Agudelo A, Papageorghiou AT, Kennedy SH, Villar J. Novel biomarkers for predicting intrauterine growth restriction: a systematic review and meta-analysis. BJOG. 2013;120:681–94.
103. Diamond JR, Wu B, Agarwal N, Bowles DW, Lam ET, Werner TL, Rasmussen E, Gamelin E, Soto F, Friberg G, Sun YN, Sharma S. Pharmacokinetic drug-drug interaction study of the angiopoietin-1/angiopoietin-2-inhibiting peptibody trebananib (amg 386) and paclitaxel in patients with advanced solid tumors. Invest New Drugs. 2015;33:691–9.
104. Goede V, Schmidt T, Kimmina S, Kozian D, Augustin HG. Analysis of blood vessel maturation processes during cyclic ovarian angiogenesis. Lab Invest. 1998;78:1385–94.
105. Tsuzuki T, Okada H, Cho H, Shimoi K, Miyashiro H, Yasuda K, Kanzaki H. Divergent regulation of angiopoietin-1, angiopoietin-2, and vascular endothelial growth factor by hypoxia and female sex steroids in human endometrial stromal cells. Eur J Obstet Gynecol Reprod Biol. 2013;168:95–101.
106. Albrecht ED, Pepe GJ. Estrogen regulation of placental angiogenesis and fetal ovarian development during primate pregnancy. Int J Dev Biol. 2010;54:397–408.

107. Harfouche R, Echavarria R, Rabbani SA, Arakelian A, Hussein MA, Hussain SN. Estradiol-dependent regulation of angiopoietin expression in breast cancer cells. J Steroid Biochem Mol Biol. 2011;123:17–24.
108. Bonagura TW, Aberdeen GW, Babischkin JS, Koos RD, Pepe GJ, Albrecht ED. Divergent regulation of angiopoietin-1 and -2, tie-2, and thrombospondin-1 expression by estrogen in the baboon endometrium. Mol Reprod Dev. 2010;77:430–8.
109. Guo B, Wang W, Li SJ, Han YS, Zhang L, Zhang XM, Liu JX, Yue ZP. Differential expression and regulation of angiopoietin-2 in mouse uterus during preimplantation period. Anat Rec (Hoboken). 2011;295:338–46.
110. Yang J, Jubb AM, Pike L, Buffa FM, Turley H, Baban D, Leek R, Gatter KC, Ragoussis J, Harris AL. The histone demethylase jmjd2b is regulated by estrogen receptor alpha and hypoxia, and is a key mediator of estrogen induced growth. Cancer Res. 2010;70:6456–66.
111. Kazi AA, Molitoris KH, Koos RD. Estrogen rapidly activates the pi3k/akt pathway and hypoxia-inducible factor 1 and induces vascular endothelial growth factor a expression in luminal epithelial cells of the rat uterus. Biol Reprod. 2009;81:378–87.
112. Molitoris KH, Kazi AA, Koos RD. Inhibition of oxygen-induced hypoxia-inducible factor-1alpha degradation unmasks estradiol induction of vascular endothelial growth factor expression in ecc-1 cancer cells in vitro. Endocrinology. 2009;150:5405–14.
113. Burger JA, Kipps TJ. Cxcr4: a key receptor in the crosstalk between tumor cells and their microenvironment. Blood. 2006;107:1761–7.
114. Liekens S, Schols D, Hatse S. Cxcl12-cxcr4 axis in angiogenesis, metastasis and stem cell mobilization. Curr Pharm Des. 2010;16:3903–20.
115. Nishigaki A, Okada H, Okamoto R, Sugiyama S, Miyazaki K, Yasuda K, Kanzaki H. Concentrations of stromal cell-derived factor-1 and vascular endothelial growth factor in relation to the diameter of human follicles. Fertil Steril. 2011;95:742–6.
116. Salcedo R, Wasserman K, Young HA, Grimm MC, Howard OM, Anver MR, Kleinman HK, Murphy WJ, Oppenheim JJ. Vascular endothelial growth factor and basic fibroblast growth factor induce expression of cxcr4 on human endothelial cells: in vivo neovascularization induced by stromal-derived factor-1alpha. Am J Pathol. 1999;154:1125–35.
117. Bachelder RE, Wendt MA, Mercurio AM. Vascular endothelial growth factor promotes breast carcinoma invasion in an autocrine manner by regulating the chemokine receptor cxcr4. Cancer Res. 2002;62:7203–6.
118. Wu X, Jin LP, Yuan MM, Zhu Y, Wang MY, Li DJ. Human first-trimester trophoblast cells recruit cd56brightcd16- nk cells into decidua by way of expressing and secreting of cxcl12/stromal cell-derived factor 1. J Immunol. 2005;175:61–8.
119. Tsutsumi A, Okada H, Nakamoto T, Okamoto R, Yasuda K, Kanzaki H. Estrogen induces stromal cell-derived factor 1 (sdf-1/cxcl12) production in human endometrial stromal cells: a possible role of endometrial epithelial cell growth. Fertil Steril. 2011;95:444–7.
120. Takenaga K. Angiogenic signaling aberrantly induced by tumor hypoxia. Front Biosci (Landmark Ed). 2011;16:31–48.
121. Schioppa T, Uranchimeg B, Saccani A, Biswas SK, Doni A, Rapisarda A, Bernasconi S, Saccani S, Nebuloni M, Vago L, Mantovani A, Melillo G, Sica A. Regulation of the chemokine receptor cxcr4 by hypoxia. J Exp Med. 2003;198:1391–402.
122. Lukiw WJ, Ottlecz A, Lambrou G, Grueninger M, Finley J, Thompson HW, Bazan NG. Coordinate activation of hif-1 and nf-kappab DNA binding and cox-2 and vegf expression in retinal cells by hypoxia. Invest Ophthalmol Vis Sci. 2003;44:4163–70.
123. Martin G, Andriamanalijaona R, Grassel S, Dreier R, Mathy-Hartert M, Bogdanowicz P, Boumediene K, Henrotin Y, Bruckner P, Pujol JP. Effect of hypoxia and reoxygenation on gene expression and response to interleukin-1 in cultured articular chondrocytes. Arthritis Rheum. 2004;50:3549–60.
124. Kishimoto H, Wang Z, Bhat-Nakshatri P, Chang D, Clarke R, Nakshatri H. The p160 family coactivators regulate breast cancer cell proliferation and invasion through autocrine/paracrine activity of sdf-1alpha/cxcl12. Carcinogenesis. 2005;26:1706–15.

125. Strydom DJ. The angiogenins. Cell Mol Life Sci. 1998;54:811–24.
126. Koga K, Osuga Y, Tsutsumi O, Yano T, Yoshino O, Takai Y, Matsumi H, Hiroi H, Kugu K, Momoeda M, Fujiwara T, Taketani Y. Demonstration of angiogenin in human endometrium and its enhanced expression in endometrial tissues in the secretory phase and the decidua. J Clin Endocrinol Metab. 2001;86:5609–14.
127. Soufla G, Sifakis S, Spandidos DA. Fgf2 transcript levels are positively correlated with egf and igf-1 in the malignant endometrium. Cancer Lett. 2008;259:146–55.
128. Maruotti N, Cantatore FP, Crivellato E, Vacca A, Ribatti D. Angiogenesis in rheumatoid arthritis. Histol Histopathol. 2006;21:557–66.
129. Shi N, Chen SY. Mechanisms simultaneously regulate smooth muscle proliferation and differentiation. J Biomed Res. 2014;28:40–6.
130. Betsholtz C. Insight into the physiological functions of pdgf through genetic studies in mice. Cytokine Growth Factor Rev. 2004;15:215–28.
131. Lindahl P, Johansson BR, Leveen P, Betsholtz C. Pericyte loss and microaneurysm formation in pdgf-b-deficient mice. Science. 1997;277:242–5.
132. Lash GE, Innes BA, Drury JA, Robson SC, Quenby S, Bulmer JN. Localization of angiogenic growth factors and their receptors in the human endometrium throughout the menstrual cycle and in recurrent miscarriage. Hum Reprod. 2012;27:183–95.
133. Dai H, Zhao S, Xu L, Chen A, Dai S. Expression of efp, vegf and bfgf in normal, hyperplastic and malignant endometrial tissue. Oncol Rep. 2010;23:795–9.
134. Evans J, Catalano RD, Brown P, Sherwin R, Critchley HO, Fazleabas AT, Jabbour HN. Prokineticin 1 mediates fetal-maternal dialogue regulating endometrial leukemia inhibitory factor. FASEB J. 2009;23:2165–75.
135. Ngan ES, Lee KY, Yeung WS, Ngan HY, Ng EH, Ho PC. Endocrine gland-derived vascular endothelial growth factor is expressed in human peri-implantation endometrium, but not in endometrial carcinoma. Endocrinology. 2006;147:88–95.
136. Battersby S, Critchley HO, Morgan K, Millar RP, Jabbour HN. Expression and regulation of the prokineticins (endocrine gland-derived vascular endothelial growth factor and bv8) and their receptors in the human endometrium across the menstrual cycle. J Clin Endocrinol Metab. 2004;89:2463–9.
137. Becker CM, Sampson DA, Short SM, Javaherian K, Folkman J, D'Amato RJ. Short synthetic endostatin peptides inhibit endothelial migration in vitro and endometriosis in a mouse model. Fertil Steril. 2006;85:71–7.
138. Cui R, Takahashi K, Takahashi F, Tanabe KK, Fukuchi Y. Endostatin gene transfer in murine lung carcinoma cells induces vascular endothelial growth factor secretion resulting in up-regulation of in vivo tumorigenecity. Cancer Lett. 2006;232:262–71.
139. O'Reilly MS, Boehm T, Shing Y, Fukai N, Vasios G, Lane WS, Flynn E, Birkhead JR, Olsen BR, Folkman J. Endostatin: an endogenous inhibitor of angiogenesis and tumor growth. Cell. 1997;88:277–85.
140. Fu GF, Li X, Hou YY, Fan YR, Liu WH, Xu GX. Bifidobacterium longum as an oral delivery system of endostatin for gene therapy on solid liver cancer. Cancer Gene Ther. 2005;12:133–40.
141. Pollheimer J, Bauer S, Huber A, Husslein P, Aplin JD, Knofler M. Expression pattern of collagen xviii and its cleavage product, the angiogenesis inhibitor endostatin, at the fetal-maternal interface. Placenta. 2004;25:770–9.
142. Nasu K, Nishida M, Fukuda J, Kawano Y, Nishida Y, Miyakawa I. Hypoxia simultaneously inhibits endostatin production and stimulates vascular endothelial growth factor production by cultured human endometrial stromal cells. Fertil Steril. 2004;82:756–9.
143. Bass R, Fernandez AM, Ellis V. Maspin inhibits cell migration in the absence of protease inhibitory activity. J Biol Chem. 2002;277:46845–8.
144. Abraham S, Zhang W, Greenberg N, Zhang M. Maspin functions as tumor suppressor by increasing cell adhesion to extracellular matrix in prostate tumor cells. J Urol. 2003;169:1157–61.

145. Huang Y, Cai LW, Yang R. Expression of maspin in the early pregnant mouse endometrium and its role during embryonic implantation. Comp Med. 2012;62:179–84.
146. Zaslavsky A, Baek KH, Lynch RC, Short S, Grillo J, Folkman J, Italiano Jr JE, Ryeom S. Platelet-derived thrombospondin-1 is a critical negative regulator and potential biomarker of angiogenesis. Blood. 2010;115:4605–13.
147. Nakamura DS, Edwards AK, Ahn SH, Thomas R, Tayade C. Compatibility of a novel thrombospondin-1 analog with fertility and pregnancy in a xenograft mouse model of endometriosis. PLoS One. 2015;10, e0121545.
148. Seki N, Kodama J, Hashimoto I, Hongo A, Yoshinouchi M, Kudo T. Thrombospondin-1 and -2 messenger rna expression in normal and neoplastic endometrial tissues: correlation with angiogenesis and prognosis. Int J Oncol. 2001;19:305–10.
149. Tan XJ, Lang JH, Zheng WM, Leng JH, Zhu L. Ovarian steroid hormones differentially regulate thrombospondin-1 expression in cultured endometrial stromal cells: implications for endometriosis. Fertil Steril. 2010;93:328–31.

Chapter 7
Oxidative Stress and Its Implications in Endometrial Function

Takeshi Kajihara, Osamu Ishihara, and Jan J. Brosens

Abstract Oxidative stress (OS) is caused by an imbalance between the production of or exposure to reactive oxygen species (ROS) and the activity of cellular antioxidants. Mammalian cells possess multiple mechanisms to remove ROS, including the utilization of enzymatic and nonenzymatic dietary antioxidants. ROS have both physiological and pathological roles in the reproductive tract. They are key signal molecules modulating various physiological reproductive functions. However, OS is associated with a number of pregnancy complications, including preeclampsia, and reproductive disorders such as endometriosis, polycystic ovary syndrome (PCOS), and unexplained infertility. The fetomaternal interface, consisting of the maternal decidua and invading placental trophoblasts, is exposed to profound changes in oxygen tension during pregnancy. These changes at the uteroplacental interface induce a burst of intracellular ROS. Endometrial decidualization is crucial to the formation of a functional fetomaternal interface because it controls endovascular trophoblast invasion and tissue homeostasis and confers resistance to environmental stress signals, including OS. Decidualizing cells surround and encapsulate the early conceptus and their remarkable resistance to oxidative cell death ensures that the pregnancy is protected against environmental stressors. In this chapter, we discuss ROS scavenger systems in the endometrium as well as the role of OS in the pathogenesis of pregnancy complications and reproductive disorders.

Keywords Human endometrium • Oxidative stress • Decidualization

T. Kajihara (✉) • O. Ishihara
Department of Obstetrics and Gynecology, Saitama Medical University, 38 Morohongo, Moroyama, Iruma-gun, Saitama 350-0495, Japan
e-mail: kajihara@saitama-med.ac.jp

J.J. Brosens
Division of Reproductive Health, Warwick Medical School, Clinical Sciences Research Laboratories, University Hospital, Coventry CV2 2DX, UK

H. Kanzaki (ed.), *Uterine Endometrial Function*,
DOI 10.1007/978-4-431-55972-6_7

7.1 Introduction

The human endometrium is a major target for ovarian sex steroid hormones. Proliferation and differentiation of the endometrium depends on the rise and fall of circulating steroid hormones during normal ovulatory cycles. In particular, the postovulatory rise in ovarian progesterone induces profound remodeling of the estrogen-primed endometrium. This remodeling is characterized initially by significant growth and coiling of the spiral arteries, secretory transformation of the glands, and then decidualization of the stromal compartment. Successful implantation depends on the dialogue between a developmentally normal embryo and a receptive endometrium. Endometrial receptivity occurs during a limited period in the menstrual cycle, designated as the "implantation window," when the endometrium allows the implantation of a blastocyst. The endometrium becomes receptive approximately 6 days after ovulation and remains so for 2–4 days [1–3]. Once the luminal epithelium is breached, the implantation embryo is rapidly embedded in the decidualizing stroma [4].

Decidualization denotes the transformation of endometrial stromal cells into specialized secretory cells, a process that is further characterized by the influx of specialized immune cells into the stroma, predominantly uterine natural killer cells and macrophages, and also by intense vascular remodeling [5]. The decidual process is crucial for the formation of a functional fetomaternal interface because it ensures tissue homeostasis during endovascular trophoblast invasion and confers resistance to environmental stress signals, including oxidative stress (OS) [6]. Failure of the endometrium to express a receptive phenotype is a major cause of conception delay and failure of in vitro fertilization (IVF) treatments. In addition, perturbations in the maternal decidual response inevitably lead to pregnancy complications [7]. Besides ovarian estrogen and progesterone, endometrial autocrine or paracrine factors and embryo-derived signals are important for implantation, although how these signals convert into a coordinated endometrial response is not well understood.

Reactive oxygen species (ROS) play a dual role in biological systems. When the balance between oxidant and antioxidant activity is maintained, ROS are known to serve as a second messenger in intracellular signal transduction cascades for various physiological cellular processes. On the other hand, excessive production of ROS can cause detrimental effects on lipids, proteins, and nucleic acids and is associated with a number of diseases. ROS also have both physiological and pathological roles in the reproductive tract. They are key signal molecules modulating various reproductive functions such as oocyte maturation, folliculogenesis, tubal function, ovulation, luteal function, and endometrial function [8]. Conversely, OS, a state characterized by an imbalance between prooxidant molecules, including ROS and reactive nitrogen species, and antioxidant defenses, plays a key role in a number of pregnancy complications such as spontaneous abortion, recurrent miscarriage, and preeclampsia [9]. Furthermore, reproductive disorders such as endometriosis, polycystic ovary syndrome (PCOS), and unexplained infertility can also develop in response to OS [8].

7.2 Reactive Oxygen Species

ROS are generated during the crucial processes of oxygen (O_2) consumption [10]. The superoxide anion radical (O^{2*-}), hydrogen peroxide (H_2O_2), and the hydroxyl radical (OH^*) are all highly reactive, diffusible, and ubiquitous molecules, which are generated as inevitable by-products of aerobic respiration and metabolism. However, the most potent and reactive ROS is superoxide, which is formed by a single-electron reduction of molecular oxygen: $O_2 + e^- \rightarrow O^{2*-}$. Hydrogen peroxide is formed on additional reduction of oxygen as follows: $2O_2^- + 2H^+ \rightarrow H_2O_2 + O_2$ [11]. Further reduction leads to the formation of OH, particularly in the presence of metal ions through the Fenton or Haber-Weiss reactions. Hydroxyl radicals are extremely reactive with a short half-life. In neutrophils, myeloperoxidase catalyzes the formation of hypochlorous acid (HClO), whereas superoxide may also react with nitric oxide (NO) to form another reactive molecule, peroxynitrite ($ONOO^-$): $O^{2*-} + NO \rightarrow ONOO^-$. The formation of superoxide anions can trigger a cascade of production of ROS, which function as key signaling molecules but can also be extremely detrimental to cells [12].

Mammalian cells possess multiple mechanisms to remove ROS, including the utilization of enzymatic and nonenzymatic dietary antioxidants. Superoxide is detoxified by superoxide dismutase (SOD) enzymes, which convert it to hydrogen peroxide. Catalase and glutathione peroxidase (GPx) further degrade hydrogen peroxidase to produce water as the end product [13] (Fig. 7.1). Physiological levels of ROS are required to ensure proper functioning of different biological pathways and to maintain homeostasis in the human body. However, excessive levels of ROS can have detrimental effects. For example, ROS can cause damage to DNA and proteins and can interfere with lipid peroxidation, which primarily affects membrane structure and function [14–16]. Therefore, a tightly regulated balance

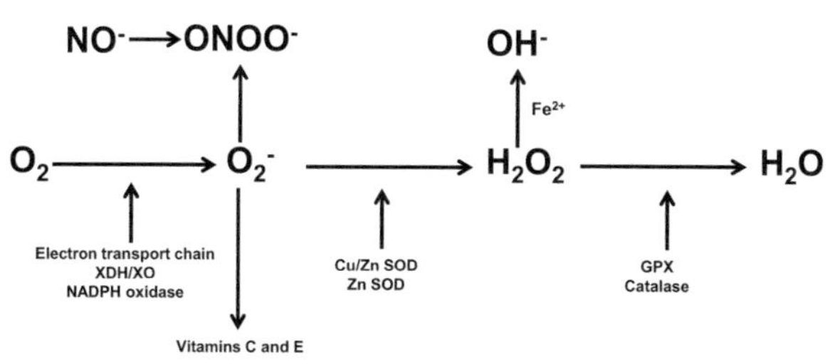

Fig. 7.1 Summary of reactive oxygen species (ROS) production and elimination. Excessive production of superoxide anions (O_2^-) can lead to the formation of hydroxyl (OH) ions through the iron-catalyzed Fenton reaction. Alternatively, O_2^- may react with nitric oxide (NO) to form the prooxidant peroxynitrite ($ONOO^-$)

between ROS and production of scavenging enzymes is required, even under physiological aerobic conditions [17].

7.3 Molecular Mechanism of Endometrial Decidualization

The most striking aspect of the decidual process is the dramatic transformation of endometrial stromal fibroblasts into secretory, epithelioid, decidual cells [18] (Fig. 7.2). In contrast to many species, decidualization of the endometrial stroma in humans is independent of the presence of an implanting blastocyst. Decidualization is first apparent in the stromal cells surrounding the terminal spiral arteries of the superficial endometrial tissue around day 23 of a 28-day cycle. In pregnancy, the decidual reaction will extend to the basal endometrial layer and critically regulates trophoblast invasion and placental formation [19]. The decidual process is characterized by the expression of a variety of phenotypic markers, including prolactin (PRL), WNT4, and insulin-like growth factor-binding protein-1 (IGFBP-1) [20]. Not surprisingly, impairment of the decidualizing process is increasingly linked to a variety of pregnancy disorders, including infertility,

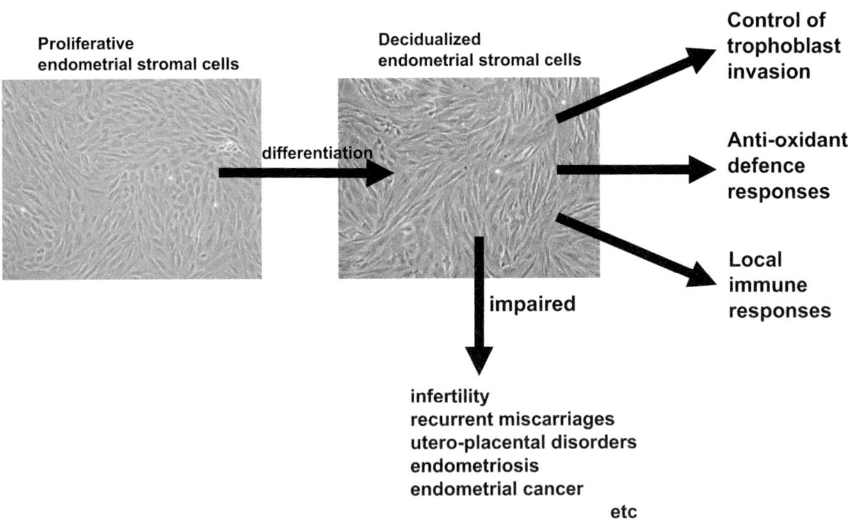

Fig. 7.2 Decidual transformation of human endometrial stromal cells (HESCs) in vitro. Undifferentiated primary HESCs display a fibroblastic spindle-shaped morphology (*left panel*). Treatment of confluent monolayers with 8-bromo-cAMP and progestin for 96 h transforms the spindle-shaped cells into cells with larger nuclei and abundant cytoplasm, resembling decidual cells (*right panel*). This transformation in vivo underpins the acquisition of specialized functions. Impairment of the decidualizing process is increasingly linked to a variety of pregnancy disorders (From Ref. [18])

recurrent miscarriages, uteroplacental disorders, endometriosis, and endometrial cancer [21–25].

Abundant clinical and experimental evidence suggests that progesterone is critical for maintaining the decidual phenotype of the endometrium prior to menstruation and throughout pregnancy. As mentioned, decidual transformation is initiated approximately 10 days after the postovulatory rise in progesterone levels [26], indicating that the expression of decidua-specific genes is unlikely to be under the direct transcriptional control of the activated progesterone receptor (PR). Furthermore, progesterone itself is a very weak inducer of the decidual phenotype in cultured primary human endometrial stromal cells (HESCs). There is compelling evidence indicating that the initiation of the decidual process requires elevated intracellular cyclic adenosine monophosphate (cAMP) levels and sustained activation of the protein kinase A (PKA) pathway [20, 27]. A ubiquitous second messenger molecule, cAMP, is generated by the binding of ligands to G protein-coupled receptors (GPCRs), also known as seven-transmembrane domain receptors. A major downstream recipient of cAMP is the cAMP-dependent PKA, a cytoplasmic enzyme composed of two regulatory and two catalytic subunits in its basal state [28]. These subunits dissociate upon cAMP binding and the catalytic subunits translocate to the nucleus, where they phosphorylate and activate cAMP-responsive element binding protein (CREB) and various cAMP-responsive element modulators (CREMs) [29]. In addition, the cAMP signaling pathway is tightly controlled at many levels, including by the expression of various receptors coupled to the PKA pathway, the degree of catabolism of cAMP by phosphodiesterases, the varied composition of the PKA holoenzyme, the expression of activating or repressing CREM isoforms, and the cell-specific composition of coactivators or corepressors.

After ovulation, circulating luteinizing hormone (LH) and follicle-stimulating hormone (FSH) levels fall, but paracrine/autocrine signals in the endometrium, such as the prostaglandin E2, relaxin, or corticotropin-releasing hormone (CRH), ensure that cAMP levels in endometrial cells continue to rise, thus triggering a decidual response [30–33]. In the case of successful implantation, the trophoblast-derived human chorionic gonadotropin (hCG) has been suggested to maintain cAMP stimulation in the decidua, although this has been comprehensively refuted [34].

Sustained elevated cAMP levels in HESCs induce the expression or trigger the activation of several downstream transcription factors, including forkhead transcription factor FOXO1, CCAAT/enhancer-binding protein β (C/EBP β), signal transducers and activators of transcription 5 (STAT5), and HOXA10. Strikingly, these decidua-specific transcription factors all physically interact with the PR, thus modulating the transcriptional competence of this nuclear receptor [6, 35–38]. The promoters of decidual *PRL* and *IGFBP1* genes are activated by multimeric transcription factor complexes that assemble in response to an interplay of cAMP- and progesterone-dependent signals [20].

7.4 Oxidative Stress and Decidualization

In early pregnancy, endovascular trophoblasts invade and effectively occlude the terminal decidual spiral arteries, thereby restricting the exposure of the early placenta and fetus to maternal arterial blood [39, 40]. Subsequently, the trophoblast plugs are dislocated, allowing perfusion of the placental intervillous space. The endovascular trophoblasts continue to invade the spiral arteries and replace the endothelial lining and most of the musculoelastic tissue in the vessel wall. This process, termed physiological transformation of the spiral arteries, ultimately establishes a high-flow, low-resistance, uteroplacental circulation, thus ensuring adequate supply to meet the increasing fetal demand for nutrients and oxygen in later pregnancy [19, 41]. As a consequence of these vascular adaptations, major fluctuations in oxygen concentrations occur at the fetomaternal interface during normal pregnancy. The partial oxygen pressure measured within the human placenta between 7 and 10 weeks of gestation is less than 20 mmHg [9]. This rises sharply to greater than 50 mmHg when the maternal perfusion of placenta is fully established, a process completed between 11 and 14 weeks of gestation. The dramatic changes in oxygen tension at the uteroplacental interface induce a burst of intracellular ROS production [9]. This increase in the levels of ROS can lead to OS when the exposure to free radicals exceeds the cellular antioxidative capacity and leads to damage to proteins, nucleic acids, and cell membranes, ultimately culminating in cell apoptosis or tissue necrosis [42]. It is increasingly recognized that oxidative damage to the placental-decidual interface represents a common pathological pathway in a spectrum of pregnancy disorders, including implantation failure, recurrent miscarriage, preeclampsia, and fetal growth restriction. To date, oxidative defense responses in the placenta have been studied extensively, but the contribution of the decidua is often ignored [5].

Sugino et al. [43] reported that the glandular epithelium showed a constant immunostaining of copper, zinc (Cu, Zn)-SOD and manganese (Mn)-SOD throughout the menstrual cycle. Stromal cells showed no apparent immunostaining for Cu, Zn-SOD or Mn-SOD during the proliferative phase, but weak to moderate immunostaining for Cu, Zn-SOD and moderate immunostaining for Mn-SOD were detected in the stromal cells showing morphologically decidualized changes in the mid-secretory phase. Furthermore, intensive immunostaining for Cu, Zn-SOD and Mn-SOD was observed in decidualized stromal cells in early pregnancy [43].

We demonstrated that decidualizing cells are remarkably resistant to oxidative cell death compared with undifferentiated HESCs [6]. This resistance during the differentiation process is partly accounted for by the induction of various free radical scavengers, most notably SOD2, monoamine oxidases A and B, thioredoxin, glutaredoxin, and peroxiredoxin. However, the exposure of undifferentiated HESCs to exogenous free radicals also rapidly activates the c-Jun NH-terminal kinase (JNK) stress signaling pathway and induces FOXO3a expression, which in turn triggers cell death. Silencing of this proapoptotic transcription factor is sufficient to prevent ROS-dependent apoptosis in HESCs [6, 44]. Massive upregulation of specific phosphatases, such as mitogen-activated protein (MAP) kinase phosphatase-1 (MKP-1), in decidualizing HESCs not only silences the JNK pathway but inhibits FOXO3a expression. Hence, undifferentiated HESCs are programmed to self-destruct in response to OS signals, whereas this mechanism is firmly disabled upon differentiation into decidual cells. Consequently, the burst of free radicals at the onset of placental perfusion could be viewed as an important stress test. If the decidual response is inadequate, the pregnancy will fail and the mother is safeguarded against the potentially catastrophic consequences of loss in advanced pregnancy. Conversely, once the stress test has been passed successfully and placental perfusion has been established around 12–14 weeks of pregnancy, the incidence of miscarriage drops dramatically.

The serum- and glucocorticoid-inducible kinase-1 (SGK1) is expressed ubiquitously and upregulated in response to cellular stress, hormones, and a myriad of other mediators. Progesterone drives the expression of SGK1 in the endometrium, first in epithelial cells and then in the decidualizing stroma cells [45–47]. SGK1 is a key regulator of sodium transport in mammalian epithelia, most prominently through its ability to directly activate the epithelial sodium channel (ENaC) and to enhance ENaC expression by inhibiting the ubiquitin ligase NEDD4-2 [48, 49]. SGK1 is also involved in proliferation and cell survival responses [50, 51]. Salker et al. [52] demonstrated that continuous SGK activity in the endometrial surface epithelium selectively disrupts the expression of implantation genes and perturbs the local fluid environment, leading to complete infertility. In pregnancy, endometrial SGK1 activity safeguards the decidual-placental interface against OS signals generated in response to intense tissue remodeling, influx of inflammatory cells, and dynamic changes in local perfusion and oxygen tension [52].

A very early event in the decidual process is a burst in endogenous free radical production, generated by cAMP-dependent activation of the nicotinamide adenine dinucleotide phosphate (NADPH) oxidase complex NOX-4/p22(PHOX), which in turn promotes the binding of C/EBP β to a core sequence within MER20, a transposable element (transposon) embedded in the promoter and enhancer regions of numerous decidual genes [53]. C/EBP β has been shown to bind FOXO1, and the recruitment of FOXO1 to MER20 converts HOXA11 from a repressor to a strong activator of the promoter. Intriguingly, the incorporation of MER20 was not the only evolutionary event that enabled maternal signaling pathways to control the decidual response. Specific evolutionary changes in the amino-acid structure of C/EBP β are thought to have also been critical in recruiting this transcription factor in the cAMP/PKA pathway [54].

It has been reported that endothelial and inducible nitric oxide synthases (eNOS and iNOS) are expressed in the human endometrium [55] and in endometrial vessels [56], and eNOS is also present in glandular epithelial cells in the human endometrium. Furthermore, NO regulates the microvasculature of the endometrium and is important for the phenomenon of menstruation. Expression of eNOS mRNA has also been identified in the mid-secretory phase and late-secretory phase, suggesting a role of NO in the decidualization of the endometrium [57].

7.5 Menstruation and Oxidative Stress

Menstruation, defined as cyclic bleeding caused by shedding of the superficial endometrium, is confined to those species that exhibit spontaneous decidualization. This breakdown of the endometrium in response to falling ovarian progesterone levels is a complex process, characterized by local leukocyte infiltration, expression and activation of matrix metalloproteinases, and apoptosis. We demonstrated that the withdrawal of progesterone from decidualized HESC cultures triggers rapid nuclear accumulation of FOXO1, which in turn activates proapoptotic genes, such as BIM, a member of the BCL-2 protein family [35]. Thus, activation of this proapoptotic pathway in decidual cells in response to falling circulating progesterone levels may be one mechanism that causes menstrual shedding of the endometrium in the absence of pregnancy [35]. On the other hand, Sugino [17] reported that when pregnancy does not occur, decreasing progesterone levels result in a decrease in Cu, Zn-SOD expression in HESCs, which in turn stimulates prostaglandin F2α (PGF2α) production via ROS. PGF2α produced by embryonic stem cells (ESCs) causes endometrial shedding via vasoconstriction. These findings suggest that ROS and SOD also play an important role in menstruation.

7.6 Endometrial Thickness and Oxidative Stress

Endometrial thickness can easily be measured by transvaginal ultrasonography. Relatively thin endometrium is associated with lower IVF success rates, although several studies that investigated the relationship between endometrial thickness and IVF outcomes produced conflicting results. Some studies demonstrated that thin endometrium negatively affects pregnancy rates [58, 59], whereas other studies could not confirm this finding [60–63]. However, a recent systematic review and meta-analysis demonstrated that the probability of clinical pregnancy in IVF recipients with an endometrial thickness <7 mm was significantly lower than those with an endometrial thickness >7 mm (23.3 % versus 48.1 %, OR 0.42, 95 % CI 0.09–0.67) [64]. It is unclear why a thin endometrium is associated with poor IVF outcomes. Unfortunately, no studies have investigated the endometrial histology of women with a thin endometrium. However, it has been speculated that when the thickness measured by ultrasound is <7 mm, the functional layer of the endometrium is thin, and the implanting embryo would be much closer to uterine spiral arteries where the higher vascularity and oxygen tension could be detrimental than the typically low oxygen tension at the surface of the endometrium. High oxygen levels may adversely affect embryo implantation [65]. In future research, the morphological characteristics of the endometrium could be studied in more detail, and molecular tools can be used to clarify the relationship between thin endometrium and OS.

7.7 Miscarriage and Oxidative Stress

Compared with other mammals, reproductive efficiency in humans is not impressive. The probability of achieving a pregnancy within one menstrual cycle is only 20–30 % [66]. Fetal chromosomal abnormalities account for approximately 50 % of all spontaneous abortions. It has been a common practice to attribute chromosomally normal pregnancy failure to a variety of maternal causes, including uterine anomalies, infection, endocrine and prothrombotic disorders, and a variety of immune abnormalities [67]. The problem with this approach is twofold: First, no identifiable cause is found in the majority of women suffering recurrent miscarriages, in spite of extensive testing. Second, many disorders perceived to cause miscarriages are also common in women with normal pregnancies, indicating a lack of causality. The miscarriage paradigm is changing based on the observation that the endometrium is an intrinsic biosensor of embryo quality. Microarray studies showed that the pregnant bovine endometrium mounts a transcriptional response that is distinct for embryos generated by artificial insemination, IVF, or somatic cell nuclear transfer [68]. Decidualizing human endometrial stromal cells have since emerged as exquisite sensors that respond to as yet ill-defined embryonic signals in a manner that either supports further development (positive selection) or ensures rapid disposal through menstruation-like shedding (negative selection) [69]. Quality control does not cease once the conceptus is embedded in the endometrium but continues throughout the first trimester of pregnancy. For example, the gradual shift from ovarian to placental progesterone production means that the endometrium will de facto select against embryos that are perceived to lack fitness because of insufficient hCG production. As already mentioned, the onset of placental perfusion around week 10 of pregnancy causes dramatic changes in local oxygen tension and triggers bursts of free radicals, effectively stress testing the fetomaternal interface.

Emerging evidence indicates that abnormal decidualization causes loss of the selectivity checkpoint, which renders the endometrium excessively permissive to implantation but unable to sustain the conceptus [52, 70]. There is considerable evidence implicating placental OS in effecting the breakdown of the fetomaternal interface. Placental trophoblasts have low concentrations and activities of antioxidant enzymes during the first trimester, rendering them particularly susceptible to OS-mediated damage. When OS develops too early in pregnancy, it can impair placentation and/or induce the degeneration of syncytiotrophoblast cells, culminating in miscarriage [67]. In addition, the decreased expression of SOD and increased levels of lipid peroxidation have been demonstrated in the decidual tissues of women suffering early pregnancy loss [71]. It is also well established that OS can also affect homeostasis in the endoplasmic reticulum (ER). Interestingly, Liu et al. [72] demonstrated that the persistence of endoplasmic OS could further sustain ER stress, eventually inducing the apoptosis of decidual cells and leading to early pregnancy loss.

7.8 Preeclampsia and Oxidative Stress

Preeclampsia is a clinical syndrome that manifests during the second half of pregnancy and affects 3–7 % of all pregnancies. Preeclampsia is one of the leading causes of maternal mortality and morbidity, particularly in developing countries [73]. This disease is clinically manifested and defined as de novo hypertension with proteinuria after 20 gestational weeks.

There is now compelling evidence that OS is present in both the maternal circulation and in the placenta in pregnancies affected by preeclampsia. Inadequate conversion of the uterine spiral arteries and subsequent impaired perfusion of the placenta provides the initiating insult [19]. In the early stages of normal pregnancy, the spiral arteries undergo substantial remodeling by surrounding interstitial trophoblasts and endovascular trophoblasts, which penetrate into the myometrium and convert muscular spiral arteries into flaccid tubes with no musculoelastic structure, capable of supplying the hugely expanded blood flow to the placenta. However, this remodeling is impaired in preeclampsia, and the decidual segments of the spiral arteries remain elastic and maintain a high resistance to blood flow. Maternal blood enters the intervillous space at a higher pressure and a faster rate, exposing placental villi to fluctuating oxygen concentrations, which contribute to ischemia-reperfusion-type injury to the placenta [74–76]. Placental ischemia-reperfusion is considered to play an important role through the induction of OS, which causes tissue damage and results in the leakage of various factors from the placenta. These factors include inflammatory cytokines, anti-angiogenic factors, and apoptotic debris, which can lead to endothelial cell dysfunction [77, 78] and systemic vasoconstriction [79]. There is a significant increase in lipid peroxidation in the placenta of preeclamptic women [80], and the activity of antioxidants, including vitamin E, vitamin C, glutathione (GSH), SOD, and thioredoxin (Trx), in the umbilical cord blood is disrupted in preeclampsia [81]. Also, the production of the antioxidant melatonin is significantly decreased in the placenta of preeclamptic women [82]. Maternal obesity has been associated with an increased risk of preeclampsia [83], which may be mediated by the increased level of OS present in obesity [84].

Recently, the use of antioxidant supplementation has attracted much attention. Some studies have reported that the vitamins C and E, as antioxidant supplementation, reduced OS in the placentas of women with preeclampsia [85–87]. However, evidence from a meta-analysis failed to find convincing evidence of beneficial effects of antioxidant supplementation to reduce the risk of preeclampsia or other complications in pregnancy [88, 89].

7.9 Other Reproductive Disorders and Oxidative Stress

Physiological levels of ROS are required to ensure proper functioning of different biological pathways and to maintain homeostasis of the human body. Low levels of free radicals act as modulators in female reproductive pathways, including oocyte maturation, physiological follicular atresia, ovulation fertilization, luteal regression, and corpus luteum formation during pregnancy. Disturbance of the physiological levels of ROS leads to female reproductive disorders, including PCOS and endometriosis. These pathologies negatively affect pregnancy rate and IVF outcome [90].

7.9.1 Endometriosis

Endometriosis is one of the most prevalent gynecological disorders. It is thought to affect approximately 10 % of women of reproductive age. Clinically, it is strongly associated with infertility and pelvic pain [91]. Endometriosis is defined as the presence of endometrial glandular and stromal cells outside the uterine cavity. While still unresolved, the most popular theory on the origin of endometriosis is the retrograde reflux of menstrual tissue via the fallopian tubes into the peritoneal cavity. Moreover, an increasing number of studies suggest that genomic, hormonal, environmental, and immunological factors and OS play important roles in the pathogenesis of endometriosis [92].

Retrograde menstruation via the fallopian tubes can cause an increase in iron levels in the female pelvis. The increased iron is taken up by macrophages, therefore reducing the ability of ferritin to remove this heavy metal. The lack of the antioxidant activity of ferritin therefore leads to OS. In turn, OS induces cellular and DNA damage and nuclear factor-kappa B (NF-κB) activation in endometriotic cells, which stimulates inflammation and cell proliferation and inhibits apoptosis, favoring the development and maintenance of endometriosis [93].

In previous studies, continuously elevated Mn-SOD and Cu, Zn-SOD, and GPx have been detected in the endometrium of women with endometriosis. Aberrant expression of these enzymes in the endometrium throughout the cycle may contribute to the establishment and development of endometriosis. Furthermore, throughout the menstrual cycle, the expression of endothelial NOS in the endometrium of women with endometriosis was persistently higher than that in the endometrium of women without endometriosis [94, 95]. However, further studies are needed to clarify the mechanisms by which the abnormal expression of these enzymes contributes to the pathophysiology of endometriosis.

It is widely accepted that endometriosis is the precursor of endometrioid and clear cell carcinomas. Several investigators have hypothesized that the iron associated with retrograde menstruation or ovarian hemorrhage in women with endometriosis participates in the pathogenesis of endometriosis-associated ovarian cancer via the generation of ROS [96–99]. Iron-dependent DNA damage caused by OS is an important factor in this carcinogenic process. Shigetomi et al. proposed that aberrant expression of AT-rich interactive domain 1A (ARID1A); phosphoinositide-3-kinase, catalytic, alpha polypeptide (PIK3CA); and NF-κB genes have been recognized as the major target genes involved in OS-induced carcinogenesis [100].

7.9.2 Polycystic Ovary Syndrome (PCOS)

One of the most common endocrine disorders, PCOS, affects approximately 6–10 % of women of reproductive age. The syndrome is associated with metabolic disorders, including insulin resistance, obesity, and diabetes. The pathogenesis of PCOS is complex and its underlying etiology remains unclear. Several characteristic features of PCOS, including androgen excess, abnormal adiposity, insulin resistance, and obesity, may contribute to the development of local and systemic OS [101, 102]. A recent systematic review and meta-analysis demonstrated that levels of several circulating markers of OS are abnormal in women with PCOS, independent of obesity [103]. These observations suggest that OS participates in the pathophysiology of PCOS. However, an association between OS and endometrial function in patients with PCOS has not yet been confirmed.

7.10 Clinical Applications

As outlined above, excessive OS and increased cell death are thought to represent a common pathological pathway in the spectrum of pregnancy disorders, from recurrent miscarriage to preeclampsia and fetal growth restriction [38, 71, 76, 104]. Consequently, there is considerable interest in using antioxidant supplements during pregnancy for the prevention of obstetrical disorders associated with impaired placental perfusion. However, several large-scale randomized trials have failed to show that vitamin C and E supplements in pregnancy are effective in preventing preeclampsia [88, 89]. The reason for the failure of antioxidant supplements to prevent obstetrical disorders is not clear but may, at least in part, reflect the importance of endogenous free radicals as signaling molecules involved in decidualization [53].

We demonstrated that hCG prevents the apoptosis of decidualizing HESCs exposed to OS in vitro [105]. Two mechanisms account for this resistance to OS-induced apoptosis. First, hCG augments SOD2 expression via FOXO1, thereby enhancing the free radical scavenging potential of decidualized HESCs. Second, hCG also antagonizes BAX expression and induces BCL-2 under OS. These observations raise the possibility that exogenous hCG treatment is useful to support early implantation events and to prevent pregnancy loss. Some clinical data support this supposition. For example, Tesarik et al. [106] reported that mid-cycle hCG administration enhanced endometrial thickness, improved implantation rates, and increased the number of multiple pregnancies in women receiving embryos from oocyte donors in oocyte donation programs. Recent data suggest that the pro-survival function of hCG in the endometrium involves the modulation of the NOTCH1 pathway [107].

Although unequivocal clinical evidence is lacking, heparin is empirically used to improve implantation [108]. A recent clinical trial demonstrated a beneficial effect of heparin, administrated during the luteal phase, on the implantation rate and the live birth rate in women with repeated implantation failure [109]. In addition to its anticoagulant activity, heparin has biological properties that could be critical for the prevention of tissue injury at the fetomaternal interface. For example, heparin suppresses natural killer cell cytotoxicity [110, 111], prevents leukocyte adhesion/influx [112–114], and antagonizes interferon-γ signaling [115]. Furthermore, Hills et al. [116] demonstrated that heparin could directly protect human villous trophoblasts against apoptosis in response to a variety of pathological stimuli. In addition, an in vitro study has shown that heparin augments PRL secretion and confers resistance against OS-induced apoptosis in decidualizing HESCs (Kajihara et al., unpublished data). These observations also raise the possibility that the administration of heparin may be useful to support early implantation events and to prevent pregnancy disorders. Clearly, well-powered clinical trials are needed to test these concepts.

7.11 Conclusions

Clearly, ROS have both physiological and pathological roles in endometrial function. The maternal decidua and invading placental trophoblasts are exposed to profound changes in oxygen tension during pregnancy. The dramatic changes in oxygen tension at the uteroplacental interface induce a burst of intracellular ROS production. Endometrial decidualized cells are remarkably resistant to oxidative cell death compared with undifferentiated HESCs. These cells have various mechanisms that confer resistance against environmental ROS. The OS resulting from an exposure to excessive levels of ROS plays a key role in a number of pregnancy complications such as spontaneous abortion, recurrent miscarriage, and preeclampsia. There is considerable interest in using antioxidant supplements during pregnancy to prevent obstetrical disorders associated with impaired placental perfusion. However, several large-scale randomized trials have found no evidence to suggest that vitamin C and E supplements in pregnancy are effective in preventing preeclampsia. We demonstrated that hCG and heparin confer resistance to OS in decidualized HESCs. Therefore, the administration of these agents may be useful to support early implantation events and thus to prevent pregnancy disorders. However, further translational studies are required to confirm the effectiveness and feasibility of these clinical applications.

References

1. Simon C, Martin JC, Pellicer A. Paracrine regulators of implantation. Baillieres Beat Pract Res Clin Obstet Gynaecol. 2000;14:815–26.
2. Strowitzki T, Germeyer A, Popvici R, von Wolff M. The human endometrium as a fertility-determining factor. Hum Reprod Update. 2006;12:617–30.
3. Bergh PA, Navot D. The impact of embryonic development and endometrial maturity on the timing of implantation. Fertil Steril. 1992;58:537–42.
4. Gellersen B, Reimann K, Samalecos A, Aupers S, Bamberger AM. Invasiveness of endometrial stromal cells is promoted by decidualization and by trophoblast-derived signals. Hum Reprod. 2010;25:862–73.
5. Gellersen B, Brosens IA, Brosens JJ. Decidualization of the human endometrium: mechanisms, functions, and clinical perspectives. Semin Reprod Med. 2007;25:445–53.
6. Kajihara T, Jones M, Fusi L, Takano M, Feroze-Zaidi F, Pirianov G, Mehmet H, Ishihara O, Higham JM, Lam EW, Brosens JJ. Differential expression of FOXO1 and FOXO3a confers resistance to oxidative cell death upon endometrial decidualization. Mol Endocrinol. 2006;20:2444–55.
7. Rai R, Regan L. Recurrent miscarriage. Lancet. 2006;368:601–11.
8. Agarwal A, Aponte-Mellado A, Premkumar BJ, Shaman A, Gupta S. The effects of oxidative stress on female reproduction: a review. Reprod Biol Endocrinol. 2012. doi:10.1186/1477-7827-10-49.
9. Jauniaux E, Watoson AL, Hempstock J, Bao YP, Skepper JN, Burton GJ. Onset of maternal arterial blood flow and placental oxidative stress. A possible factor in human early pregnancy failure. Am J Pathol. 2000;157:2111–22.
10. Fujii J, Iuchi Y, Okada F. Fundamental roles of reactive oxygen species and protective mechanisms in the female reproductive system. Reprod Biol Endocrinol. 2005;3:43.
11. Babior BM. NADPH oxidase: an update. Blood. 1999;93:1464–76.
12. Myatt SS, Brosens JJ, Lam EW. Sense and sensitivity: FOXO and ROS in cancer development and treatment. Antioxid Redox Signal. 2011;14:675–87.
13. Halliwell B, Gutteridge JMC. Free radical biology and medicine. 4th ed. Oxford: Oxford Science; 2007.
14. Fridovich I. Superoxide radical: an endogenous toxicant. Annu Rev Pharmacol Toxicol. 1983;23:239–57.
15. Slater TF. Free-radical mechanisms in tissue injury. Biochem J. 1984;222:1–15.
16. Fridovich I. Biological effects of the superoxide radical. Arch Biochem Biophys. 1986;247:1–11.
17. Sugino N. The role of oxygen radical-mediated signaling pathways in endometrial function. Placenta. 2007;28:S133–6.
18. Kajihara T, Brosens JJ, Ishihara O. The role of FOXO1 in the decidual transformation of the endometrium and early pregnancy. Med Mol Morphol. 2013;46:61–8.
19. Brosens JJ, Pijnenborg R, Brosens IA. The myometrial junctional zone spiral arteries in normal and abnormal pregnancies: a review of the literature. Am J Obstet Gynecol. 2002;187:1416–23.
20. Gellersen B, Brosens J. Cyclic AMP and progesterone receptor cross-talk in human endometrium: a decidualizing affair. J Endocrinol. 2003;178:357–72.
21. Maruyama T, Yoshimura Y. Molecular and cellular mechanisms for differentiation and regeneration of the uterine endometrium. Endocr J. 2008;55:795–810.
22. Brosens JJ, Hayashi N, White JO. Progesterone receptor regulates decidual prolactin expression in differentiating human endometrial stromal cells. Endocrinology. 1999;140:4809–20.
23. Brosens J, Brosens I. Adenomyosis. In: Cameron I, Fraser I, Smith S, editors. Clinical disorders of the menstrual cycle and endometrium. New York: Oxford University Press; 1998. p. 297–312.

24. Bulun SE, Noble LS, Takayama K, Michael MD, Agarwal V, Fisher C, Zhao Y, Hinshelwood MM, Ito Y, Simpson ER. Endocrine disorders associated with inappropriately high aromatase expression. J Steroid Biochem Mol Biol. 1997;61:133–9.
25. Daly DC, Maslar IA, Rosenberg SM, Tohan N, Riddick DH. Prolactin production by luteal phase defect endometrium. Am J Obstet Gynecol. 1981;140:587–91.
26. De Ziegler D, Fanchin R, de Moustier B, Bulletti C. The hormonal control of endometrial receptivity: estrogen (E2) and progesterone. J Reprod Immunol. 1998;39:149–66.
27. Telgmann R, Maronde E, Tasken K, Gellersen B. Activated protein kinase A is required for differentiation-dependent transcription of the decidual prolactin gene in human endometrial stromal cells. Endocrinology. 1997;138:929–37.
28. Shalhegg BS, Tasken K. Specificity in the cAMP/PKA signaling pathway, differential expression, regulation, and subcellular localization of subunits of PKA. Front Biosci. 2000;5:D678–93.
29. Mayr B, Montminy M. Transcriptional regulation by the phosphorylation-dependent factor CREB. Nat Rev Mol Cell Biol. 2001;2:599–609.
30. Tseng L, Gao JG, Chen R, Zhu HH, Mazella J, Powell DR. Effect of progestin, antiprogestin, and relaxin on the accumulation of prolactin and insulin-like growth factor-binding protein-1 messenger ribonucleic acid in human endometrial stromal cells. Biol Reprod. 1992;47:441–50.
31. Tang B, Gurpide E. Direct effect of gonadotropins on decidualization of human endometrial stromal cells. J Steroid Biochem Mol Biol. 1993;47:115–21.
32. Frank GR, Brar AK, Cedars MI, Handwerger S. Prostaglandin E2 enhances human endometrial stromal cell differentiation. Endocrinology. 1994;134:258–63.
33. Ferrari A, Petraglia F, Gurpide E. Corticotropin releasing factor decidualized human endometrial stromal cells *in vitro*. Interaction with progestin. J Steroid Biochem Mol Biol. 1995;54:251–5.
34. Banerjee P, Sapru K, Strakova Z, Fazleabas AT. Chorionic gonadotropin regulates prostaglandin E synthase via a phosphatidylinositol 3-kinase-extracellular regulatory kinase pathway in a human endometrial epithelial cell line: implications for endometrial responses for embryo implantation. Endocrinology. 2009;150:4326–37.
35. Labied S, Kajihara T, Madureira PA, Fusi L, Jones MC, Higham JM, Varshochi R, Francis JM, Zoumpoulidou G, Essafi A, Fernandez de Mattos S, Lam EW, Brosens JJ. Progestins regulate the expression and activity of the Forkhead transcription factor FOXO1 in differentiating human endometrium. Mol Endocrinol. 2006;20:35–44.
36. Christian M, Zhang X, Schneider-Merck T, Unterman TG, Gellersen B, White JO, Brosens JJ. Cyclic AMP-induced forkhead transcription factor, FKHR, cooperates with CCAAT/enhancer-binding protein β in differentiating human endometrial stromal cells. J Biol Chem. 2002;277:20825–32.
37. Christian M, Pohnke Y, Kempf R, Gellersen B, Brosens JJ. Functional association of PR and CCAAT/enhancer-binding protein β isoforms: promoter-dependent cooperation between PR-B and liver-enriched inhibitor protein, or liver-enriched activity protein and PR-A in human endometrial stromal cells. Mol Endocrinol. 2002;16:141–54.
38. Mak IYH, Brosens JJ, Christian M, Hills FA, Chamley L, Regan L, White JO. Regulated expression of signal transducer and activator of transcription, Stat5, and its enhancement of PRL expression in human endometrial stromal cells *in vitro*. J Clin Endocrinol Metab. 2002;87:2581–8.
39. Burton GJ, Jauniaux E, Watson AL. Maternal arterial connections to the placental intervillous space during the first trimester of human pregnancy: the Boyd collection revisited. Am J Obstet Gynecol. 1999;181:718–24.
40. Burton GJ, Jauniaux E. Placental oxidative stress: from miscarriage to preeclampsia. J Soc Gynecol Investig. 2004;11:342–52.
41. Brosens I, Robertson WB, Dixon HG. The physiological response of the vessels of the placental bed to normal pregnancy. J Pathol Bacteriol. 1967;93:569–79.

42. Lynch RE, Fridovich I. Effects of superoxide on the erythrocyte membrane. J Biol Chem. 1978;25:1838–45.
43. Sugino N, Shimamura K, Takiguchi S, Tamura H, Ono M, Nakata M, Nakamura Y, Ogino K, Uda T, Kato H. Changes in activity of superoxide dismutase in the human endometrium throughout the menstrual cycle and in early pregnancy. Hum Reprod. 1996;11:1073–8.
44. Leitao B, Jones MC, Fusi L, Higham J, Lee Y, Takano M, Goto T, Christian M, Lam EW, Brosens JJ. Silencing of the JNK pathway maintains progesterone receptor activity in decidualizing human endometrial stromal cells exposed to oxidative stress signals. FASEB J. 2010;24:1541–51.
45. Feroze-Zaidi F, Fusi L, Takano M, Higham J, Salker MS, Goto T, Edassery S, Klingel K, Boini KM, Palmada M, Kamps R, Groothuis PG, Lam EW, Smith SK, Lang F, Sharkey AM, Brosens JJ. Role and regulation of the serum- and glucocorticoid-regulated kinase 1 in fertile and infertile human endometrium. Endocrinology. 2007;148:5020–9.
46. Jeong JW, Lee KY, Kwak I, White LD, Hilsenbeck SG, Lydon JP, DeMayo FJ. Identification of murine uterine genes regulated in a ligand-dependent manner by the progesterone receptor. Endocrinology. 2005;146:3490–505.
47. Talbi S, Hamilton AE, Vo KC, Tulac S, Overgaard MT, Dosiou C, Le Shay N, Nezhat CN, Kempson R, Lessey BA, Nayak NR, Giudice LC. Molecular phenotyping of human endometrium distinguishes menstrual cycle phases and underlying biological processes in normo-ovulatory women. Endocrinology. 2006;147:1097–121.
48. Lang F, Böhmer C, Palmada M, Seebohm G, Strutz-Seebohm N, Vallon V. (Patho)physiological significance of the serum- and glucocorticoid-inducible kinase isoforms. Physiol Rev. 2006;86:1151–78.
49. Loffing J, Flores SY, Staub O. Sgk kinases and their role in epithelial transport. Annu Rev Physiol. 2006;68:461–90.
50. Amato R, D'Antona L, Porciatti G, Agosti V, Menniti M, Rinaldo C, Costa N, Bellacchio E, Mattarocci S, Fuiano G, Soddu S, Paggi MG, Lang F, Perrotti N. Sgk1 activates MDM2-dependent p53 degradation and affects cell proliferation, survival, and differentiation. J Mol Med (Berl). 2009;87:1221–39.
51. Brunet A, Park J, Tran H, Hu LS, Hemmings BA, Greenberg ME. Protein kinase SGK mediates survival signals by phosphorylating the forkhead transcription factor FKHRL1 (FOXO3a). Mol Cell Biol. 2001;21:952–65.
52. Salker MS, Christian M, Steel JH, Nautiyal J, Lavery S, Trew G, Webster Z, Al-Sabbagh M, Puchchakayala G, Föller M, Landles C, Sharkey AM, Quenby S, Aplin JD, Regan L, Lang F, Brosens JJ. Deregulation of the serum- and glucocorticoid-inducible kinase SGK1 in the endometrium causes reproductive failure. Nat Med. 2011;17:1509–13.
53. Al-Sabbagh M, Fusi L, Higham J, Lee Y, Lei K, Hanyaloglu AC, Lam EW, Christian M, Brosens JJ. NADPH oxidase-derived reactive oxygen species mediate decidualization of human endometrial stromal cells in response to cyclic AMP signaling. Endocrinology. 2011;152:730–40.
54. Lynch VJ, May G, Wagner GP. Regulatory evolution through divergence of a phosphoswitch in the transcription factor CEBPB. Nature. 2011;480:383–6.
55. Tseng L, Zhang J, Peresleni TY, Goligorsky MS. Cyclic expression of endothelial nitric oxide synthase mRNA in the epithelial glands of human endometrium. J Soc Gynecol Investig. 1996;3:33–8.
56. Taguchi M, Alfer J, Chwalisz K, Beier HM, Classen-Linke I. Endothelial nitric oxide synthase is differently expressed in human endometrial vessels during the menstrual cycle. Mol Hum Reprod. 2000;6:185–90.
57. Agarwal A, Gupta S, Sharma R. Oxidative stress and its implications in female infertility – a clinician's perspective. Reprod Biomed Online. 2005;11:641–50.
58. Kovacs P, Matyas S, Boda K, Kaali SG. The effect of endometrial thickness on IVF/ICSI outcome. Hum Reprod. 2003;18:2337–41.

59. El-Toukhy T, Coomarasamy A, Khairy M, Sunkara K, Seed P, Khalaf Y, Braude P. The relationship between endometrial thickness and outcome of medicated frozen embryo replacement cycles. Fertil Steril. 2008;89:832–9.
60. Oliveira JB, Baruffi RL, Mauri AL, Petersen CG, Borges MC, Franco Jr JG. Endometrial ultrasonography as a predictor of pregnancy in an in-vitro fertilization programme after ovarian stimulation and gonadotrophin-releasing hormone and gonadotrophins. Hum Reprod. 1997;12:2515–8.
61. Lesny P, Killick SR, Tetlow RL, Manton DJ, Robinson J, Maguiness SD. Ultrasound evaluation of the uterine zonal anatomy during in-vitro fertilization and embryo transfer. Hum Reprod. 1999;14:1593–8.
62. Yuval Y, Lipitz S, Dor J, Achiron R. The relationships between endometrial thickness, and blood flow and pregnancy rates in in-vitro fertilization. Hum Reprod. 1999;14:1067–71.
63. Mercé LT, Barco MJ, Bau S, Troyano J. Are endometrial parameters by three-dimensional ultrasound and power Doppler angiography related to in vitro fertilization/embryo transfer outcome? Fertil Steril. 2008;89:111–7.
64. Kasius A, Smit JG, Torrance HL, Eijkemans MJ, Mol BW, Opmeer BC, Broekmans FJ. Endometrial thickness and pregnancy rates after IVF: a systematic review and meta-analysis. Hum Reprod Update. 2014;20:530–41.
65. Casper RF. It's time to pay attention to the endometrium. Fertil Steril. 2011;96:519–21.
66. Macklon NS, Brosens JJ. The human endometrium as a sensor of embryo quality. Biol Reprod. 2014;91:98.
67. Gupta S, Agarwal A, Banerjee J, Alvarez JG. The role of oxidative stress in spontaneous abortion and recurrent pregnancy loss: a systematic review. Obstet Gynecol Surv. 2007;62:335–47.
68. Mansouri-Attia N, Sandra O, Aubert J, Degrelle S, Everts RE, Giraud-Delville C, Heyman Y, Galio L, Hue I, Yang X, Tian XC, Lewin HA, Renard JP. Endometrium as an early sensor of in vitro embryo manipulation technologies. Proc Natl Acad Sci U S A. 2009;106:5687–92.
69. Brosens JJ, Salker MS, Teklenburg G, Nautiyal J, Salter S, Lucas ES, Steel JH, Christian M, Chan YW, Boomsma CM, Moore JD, Hartshorne GM, Sućurović S, Mulac-Jericevic B, Heijnen CJ, Quenby S, Koerkamp MJ, Holstege FC, Shmygol A, Macklon NS. Uterine selection of human embryos at implantation. Sci Rep. 2014;4:3894.
70. Salker M, Teklenburg G, Molokhia M, Lavery S, Trew G, Aojanepong T, Mardon HJ, Lokugamage AU, Rai R, Landles C, Roelen BA, Quenby S, Kuijk EW, Kavelaars A, Heijnen CJ, Regan L, Macklon NS, Brosens JJ. Natural selection of human embryos: impaired decidualization of endometrium disables embryo-maternal interactions and causes recurrent pregnancy loss. PLoS One. 2010;5, e10287.
71. Sugino N, Nakata M, Kashida S, Karube A, Takiguchi S, Kato H. Decreased superoxide dismutase expression and increased concentrations of lipid peroxide and prostaglandin F (2alpha) in the decidua of failed pregnancy. Mol Hum Reprod. 2000;6:642–7.
72. Liu AX, He WH, Yin LJ, Lv PP, Zhang Y, Sheng JZ, Leung PC, Huang HF. Sustained endoplasmic reticulum stress as a cofactor of oxidative stress in decidual cells from patients with early pregnancy loss. J Clin Endocrinol Metab. 2011;96:E493–7.
73. Hansson SR, Nääv Å, Erlandsson L. Oxidative stress in preeclampsia and the role of free fetal hemoglobin. Front Physiol. 2015;5:516.
74. Jauniaux E, Jurkovic D, Campbell S. Current topic: in vivo investigation of the placental circulations by Doppler echography. Placenta. 1995;16:323–31.
75. Jauniaux E, Ramsay B, Campbell S. Ultrasonographic investigation of placental morphologic characteristics and size during the second trimester of pregnancy. Am J Obstet Gynecol. 1994;170:130–7.
76. Hung TH, Skepper JN, Burton GJ. In vitro ischemia-reperfusion injury in term human placenta as a model for oxidative stress in pathological pregnancies. Am J Pathol. 2001;159:1031–43.
77. Myatt L, Cui X. Oxidative stress in the placenta. Histochem Cell Biol. 2004;122:369–82.

78. Reslan OM, Khalil RA. Molecular and vascular targets in the pathogenesis and management of the hypertension associated with preeclampsia. Cardiovasc Hematol Agents Med Chem. 2010;8:204–26.
79. Khalil RA, Granger JP. Vascular mechanisms of increased arterial pressure in preeclampsia: lessons from animal models. Am J Physiol Regul Integr Comp Physiol. 2002;283:R29–45.
80. Vanderlelie J, Venardos K, Clifton VL, Gude NM, Clarke FM, Perkins AV. Increased biological oxidation and reduced anti-oxidant enzyme activity in pre-eclamptic placentae. Placenta. 2005;26:53–8.
81. Matsubara K, Higaki T, Matsubara Y, Nawa A. Nitric oxide and reactive oxygen species in the pathogenesis of preeclampsia. Int J Mol Sci. 2015;16:4600–14.
82. Lanoix D, Guérin P, Vaillancourt C. Placental melatonin production and melatonin receptor expression are altered in preeclampsia: new insights into the role of this hormone in pregnancy. J Pineal Res. 2012;53:417–25.
83. Wolf M, Kettyle E, Sandler L, Ecker JL, Roberts J, Thadhani R. Obesity and preeclampsia: the potential role of inflammation. Obstet Gynecol. 2001;98:757–62.
84. Zavalza-Gómez AB. Obesity and oxidative stress: a direct link to preeclampsia? Arch Gynecol Obstet. 2011;283:415–22.
85. Poston L, Briley AL, Seed PT, Kelly FJ, Shennan AH. Vitamins in pre-eclampsia (VIP) trial consortium. Vitamin C and vitamin E in pregnant women at risk for pre-eclampsia (VIP trial): randomised placebo-controlled trial. Lancet. 2006;367:1145–54.
86. Rumbold AR, Crowther CA, Haslam RR, Dekker GA, Robinson JS, ACTS Study Group. Vitamins C and E and the risks of preeclampsia and perinatal complications. N Engl J Med. 2006;354:1796–806.
87. Villar J, Purwar M, Merialdi M, Zavaleta N, Thi Nhu Ngoc N, Anthony J, De Greeff A, Poston L, Shennan A, WHO Vitamin C and Vitamin E Trial Group. World Health Organisation multicentre randomised trial of supplementation with vitamins C and E among pregnant women at high risk for pre-eclampsia in populations of low nutritional status from developing countries. BJOG. 2009;116:780–8.
88. Rumbold A, Duley L, Crowther CA, Haslam RR. Antioxidants for preventing pre-eclampsia. Cochrane Database Syst Rev. 2008;1, CD004227.
89. Conde-Agudelo A, Romero R, Kusanovic JP, Hassan SS. Supplementation with vitamins C and E during pregnancy for the prevention of preeclampsia and other adverse maternal and perinatal outcomes: a systematic review and metaanalysis. Am J Obstet Gynecol. 2011;204:503.e1–12.
90. Gupta S, Ghulmiyyah J, Sharma R, Halabi J, Agarwal A. Power of proteomics in linking oxidative stress and female infertility. Biomed Res Int. 2014;2014:916212.
91. Holoch KJ, Lessey BA. Endometriosis and infertility. Clin Obstet Gynecol. 2010;53:429–38.
92. Prechapanich J, Kajihara T, Fujita K, Sato K, Uchino S, Tanaka K, Matsumoto S, Akita M, Nagashima M, Brosens JJ, Ishihara O. Effect of a dienogest for an experimental three-dimensional endometrial culture model for endometriosis. Med Mol Morphol. 2014;47:189–95.
93. Lousse JC, Van Langendonckt A, Defrere S, Ramos RG, Colette S, Donnez J. Peritoneal endometriosis is an inflammatory disease. Front Biosci (Elite Ed). 2012;4:23–40.
94. Ota H, Igarashi S, Hatazawa J, Tanaka T. Endothelial nitric oxide synthase in the endometrium during the menstrual cycle in patients with endometriosis and adenomyosis. Fertil Steril. 1998;69:303–8.
95. Ota H, Igarashi S, Hatazawa J, Tanaka T. Immunohistochemical assessment of superoxide dismutase expression in the endometrium in endometriosis and adenomyosis. Fertil Steril. 1999;72:129–34.
96. Yamaguchi K, Mandai M, Toyokuni S, Hamanishi J, Higuchi T, Takakura K, Fujii S. Contents of endometriotic cysts, especially the high concentration of free iron, are a possible cause of carcinogenesis in the cysts through the iron-induced persistent oxidative stress. Clin Cancer Res. 2008;14:32–40.

97. Kobayashi H, Yamada Y, Kanayama S, Furukawa N, Noguchi T, Haruta S, Yoshida S, Sakata M, Sado T, Oi H. The role of iron in the pathogenesis of endometriosis. Gynecol Endocrinol. 2009;25:39–52.
98. Mandai M, Yamaguchi K, Matsumura N, Baba T, Konishi I. Ovarian cancer in endometriosis: molecular biology, pathology, and clinical management. Int J Clin Oncol. 2009;14:383–91.
99. Kobayashi H, Kajiwara H, Kanayama S, Yamada Y, Furukawa N, Noguchi T, Haruta S, Yoshida S, Sakata M, Sado T, Oi H. Molecular pathogenesis of endometriosis-associated clear cell carcinoma of the ovary (review). Oncol Rep. 2009;22:233–40.
100. Shigetomi H, Tsunemi T, Haruta S, Kajihara H, Yoshizawa Y, Tanase Y, Furukawa N, Yoshida S, Sado T, Kobayashi H. Molecular mechanisms linking endometriosis under oxidative stress with ovarian tumorigenesis and therapeutic modalities. Cancer Invest. 2012;30:473–80.
101. Vincent HK, Taylor AG. Biomarkers and potential mechanisms of obesity-induced oxidant stress in humans. Int J Obes (Lond). 2006;30:400–18.
102. Liu S, Navarro G, Mauvais-Jarvis F. Androgen excess produces systemic oxidative stress and predisposes to beta-cell failure in female mice. PLoS One. 2010;5, e11302.
103. Murri M, Luque-Ramírez M, Insenser M, Ojeda-Ojeda M, Escobar-Morreale HF. Circulating markers of oxidative stress and polycystic ovary syndrome (PCOS): a systematic review and meta-analysis. Hum Reprod Update. 2013;19:268–88.
104. Redman CW, Sargent IL. Latest advances in understanding preeclampsia. Science. 2005;308:1592–4.
105. Kajihara T, Uchino S, Suzuki M, Itakura A, Brosens JJ, Ishihara O. Human chorionic gonadotropin confers resistance to oxidative stress-induced apoptosis in decidualizing human endometrial stromal cells. Fertil Steril. 2011;95:1302–7.
106. Tesarik J, Hazout A, Medoza C. Luteinizing hormone affects uterine receptivity independently of ovarian function. Reprod Biomed Online. 2003;7:59–64.
107. Afshar Y, Miele L, Fazleabas AT. Notch1 is regulated by chorionic gonadotropin and progesterone in endometrial stromal cells and modulates decidualization in primates. Endocrinology. 2012;153:2884–96.
108. Ricci G, Giolo E, Simeone R. Heparin's 'potential to improve pregnancy rates and outcomes' is not evidence-based. Hum Reprod Update. 2010;16:225–7.
109. Urman B, Ata B, Yakin K, Alatas C, Aksoy S, Mercan R, Balaban B. Luteal phase empirical low molecular weight heparin administration in patients with failed ICSI embryo transfer cycles: a randomized open-labeled pilot trial. Hum Reprod. 2009;24:1640–7.
110. Yamamoto H, Fuyama S, Arai S, Sendo F. Inhibition of mouse natural killer cytotoxicity by heparin. Cell Immunol. 1985;96:409–17.
111. Johann S, Zoller C, Haas S, Blumel G, Lipp M, Forster R. Sulfated polysaccharide anticoagulants suppress natural killer cell activity in vitro. Thromb Haemost. 1995;74:998–1002.
112. Christopherson II KW, Campbell JJ, Travers JB, Hromas RA. Low-molecular-weight heparins inhibit CCL2l-induced T cell adhesion and migration. J Pharmacol Exp Ther. 2002;302:290–5.
113. Manduteanu I, Voinea M, Capraru M, Dragomir E, Simionescu M. A novel attribute of enoxaparin: inhibition of monocyte adhesion to endothelial cells by a mechanism involving cell adhesion molecules. Pharmacology. 2002;65:52–3.
114. Wan JG, Mu JS, Zhu HS, Geng JG. N-desulfated non-anticoagulant heparin inhibits leukocyte adhesion and transmigration in vitro and attenuates acute peritonitis and ischemia and reperfusion injury in vitro. Inflamm Res. 2002;51:435–43.
115. Fritchley SJ, Kirby JA, Ali S. The antagonism of interferon-gamma (INF-gamma) by heparin: examination of the blockade of class II MHC antigen and heat shock protein-70 expression. Clin Exp Med Biol. 1992;120:247–52.
116. Hills FA, Abrahams VM, Gonzalez-Timon B, Franvis J, Clock B, Hinkson L, Rai R, Mor Regan L, Sullivan M, Lam EW, Brosens JJ. Heparin prevents programmed cell death in human trophoblast. Mol Hum Reprod. 2006;12:237–43.

Chapter 8
Decidualization and Epigenetic Regulation

Norihiro Sugino, Isao Tamura, Ryo Maekawa, and Kosuke Jozaki

Abstract Decidualization is a process of differentiation of the endometrial stromal cells (ESC) accompanied by dramatic changes of cell functions and is necessary for embryo implantation and pregnancy establishment. A number of genes are upregulated or downregulated during decidualization. The present review focused on the involvement of epigenetics (histone modifications, DNA methylation, and microRNAs) in the regulation of gene expression in human ESC during decidualization. Histone modification statuses genome-widely change during decidualization. The main histone modifications are increases of H3K27ac and H3K4me3, which activate transcription, whereas decreases of these histone modifications are quite few. The increases of H3K27ac and H3K4me3 are observed not only in the proximal region but also in the distal promoter regions, suggesting distal enhancer-proximal promoter interactions are involved in the upregulation of gene expression during decidualization. One of the potential roles of the increases in H3K27ac and H3K4me3 during decidualization is to activate the insulin signaling pathway, which contributes to decidualization through the increase of glucose uptake. On the other hand, genome-wide DNA methylation does not remarkably change in human ESC during decidualization. These findings indicate that epigenetic regulation, especially through histone modifications, plays important roles in the decidualization of human ESC.

Keywords Endometrium • Epigenetics • DNA methylation • Decidualization • Histone modification

8.1 Introduction

Implantation of the human embryo in the maternal endometrium is a key in the establishment of pregnancy and requires a dialog between the embryo and the receptive endometrium. In human, this process is initiated by progesterone and

N. Sugino (✉) • I. Tamura • R. Maekawa • K. Jozaki
Department of Obstetrics and Gynecology, Yamaguchi University Graduate School of Medicine, Minamikogushi 1-1-1, Ube, Yamaguchi 755-8505, Japan
e-mail: sugino@yamaguchi-u.ac.jp

H. Kanzaki (ed.), *Uterine Endometrial Function*,
DOI 10.1007/978-4-431-55972-6_8

includes morphological and functional changes of the fibroblast-like stromal cells. Decidualization is the morphological and functional transformation of endometrial stromal fibroblasts into differentiated decidual cells, which is critical for embryo implantation and establishment of pregnancy. A defective decidualization response is associated with reproductive disorders such as early pregnancy loss and recurrent miscarriage [1, 2]. Endometrial stromal cells that are primed by estrogen in the proliferative phase respond to progesterone and become larger and rounder with activation of a range of functional pathways. Genome-wide microarray analyses have identified a number of genes that are upregulated or downregulated by decidualization in human endometrial stromal cells [3–5], suggesting that decidualization is a process of differentiation accompanied by dramatic changes of cell functions.

Decidual transformation involves many factors, including intracellular signaling pathways and their related molecules and transcription factors [3–6]. Recently, it has become clear that changes in chromatin structure of the promoter region play key roles in gene expression. Thus, epigenetic changes such as histone modifications and DNA methylation can control transcription by affecting the chromatin structure of the gene promoter [7, 8]. Histone acetylases and histone deacetylases (HDACs) are constitutively expressed in the human endometrium [9]. Decidualization has also been shown to be associated with histone modifications in vitro [10]. A chromatin immunoprecipitation combined with a microarray analysis (ChIP-on-chip) showed that the histone modification status around the transcription start site is altered in human endometrial stromal cells during decidualization [11]. Regarding DNA methylation, DNA methyltransferases (DNMTs) are expressed at the mRNA levels in the human endometrium and decrease during the mid-late secretory phase [12]. In addition, estrogen and progesterone treatment in human endometrial stromal cell cultures decreased DNMT3a and DNMT3b mRNA expression [12]. The DNA methylation status of many genes in the endometrium changes between the proliferative and mid-secretory phases [13]. These findings suggest that histone modifications and DNA methylation are involved in the regulation of gene expression of human endometrial stromal cells undergoing decidualization. In this chapter, we reviewed the significance of epigenetics in the regulation of gene expression in human endometrial stromal cells during decidualization.

8.2 Epigenetic Regulation

Epigenetics means “the study of mitotically and/or meiotically heritable changes in gene function that cannot be explained by changes in DNA sequence” [14]. Epigenetics is defined as a mechanism that activates or represses gene expression without detectable changes in DNA sequence, which includes histone modifications, DNA methylation, and posttranscriptional regulation by microRNAs.

8.2.1 Histone Modifications

In the eukaryotic nucleus, about 146 bp of DNA is wrapped around a histone octamer consisting of two sets of the core histones, H2A, H2B, H3, and H4, to form a nucleosome [15]. The nucleosomes are linked by loops of DNA and linker histone H1, forming chromatin. The accessibility of transcription factors to the binding site of a promoter depends strongly on the chromatin structure of the promoter [7]. The chromatin structure is regulated by histone modifications. Lysine residues in the N-terminal tails (histone tails) of histone H3 or histone H4 are modified by acetylation or methylation. Activation or repression of transcription is determined by the location and histone modification types of the lysine residue. Acetylation of histone H3 lysine-27 (H3K27ac) and monomethylation or trimethylation of histone H3 lysine-4 (H3K4me1, H3K4me3) make chromatin structure loose, which is associated with activation of gene transcription, while trimethylation of histone H3 lysine-27 (H3K27me3) makes chromatin structure condensed, which is associated with transcriptional repression [16–18]. Acetylation of any lysine residues in the histone tail activates transcription. All known histone modifications have now been shown to reversible, and some histone modifications have been shown to change within minutes of a stimulus, indicating that histone modifications are very dynamic.

8.2.2 DNA Methylation

DNA methylation occurs at cytosines within CpG dinucleotides that are frequently clustered in regions of 1–2 kb in length, called CpG islands, in or near the promoter and first exon regions of genes. In vertebrates, large un-methylated GC-rich regions can be found at the 5′ end of many genes. DNA methylation at the promoter is associated with transcriptional silencing. CpG methylation in the promoter downregulates gene expression by changing the chromatin structure and preventing the binding of transcription factors [19]. In contrast, in the hypomethylation status of the promoter region, transcriptional factors can access their binding sites on the promoter to activate transcription. Changes in genomic DNA methylation occur during the course of normal development, and epigenetic regulation of gene expression by DNA methylation is thought to be an important mechanism in cell-specific gene expression [20].

8.2.3 MicroRNAs (miRNAs)

miRNAs are endogenous, short, noncoding RNA molecules that regulate gene expression by translational repression or degradation of mRNA in a sequence-specific manner [21]. miRNAs bind to the complementary target sequences in the 3′ untranslated regions of mRNAs and direct the translational repression or degradation of target mRNAs. A growing body of evidence has indicated that miRNAs have a role in a variety of biological processes including development, differentiation, apoptosis, and cell proliferation.

8.3 Histone Modifications of the Gene Promoter and Gene Expression

We explain the significance of histone modifications in the regulation of gene expression in human endometrial stromal cells undergoing decidualization. Insulin-like growth factor binding protein-1 (IGFBP-1) is preferentially expressed by decidualization and is recognized as a specific marker of decidualization. Human endometrial stromal cells isolated from the proliferative phase endometrium were incubated with cAMP to induce decidualization. cAMP induced the expression of IGFBP-1 in human endometrial stromal cells, but it did not induce IGFBP-1 in human dermal fibroblasts (HDF), which were used as a non-endometrial control [22]. This is due to the low level of histone acetylation of the IGFBP-1 promoter region in HDF. In fact, co-treatment of HDF with cAMP and HDAC inhibitors, which increase histone acetylation, increased the histone acetylation level of the IGFBP-1 promoter region, which in turn induced the recruitment of C/EBP β, a transcription factor, to the promoter region, resulting in induction of IGFBP-1 expression [22]. These results indicate that the high histone acetylation status of the promoter region of IGFBP-1 is associated with the induction of the IGFBP-1 gene by making the promoter region accessible to transcriptional factors. Decidualization is also associated with changes in histone modifications in vitro [10, 23]. Treating human endometrial stromal cells with the HDAC inhibitor resulted in morphological changes and expressions of IGFBP-1 and prolactin, which is another marker of decidualization [10]. Histones H3 and H4 became acetylated upon decidualization and the acetylation of H4 in turn was associated with transcriptional activation in the promoter region including the progesterone receptor binding site of IGFBP-1 [10].

8.4 Genome-Wide Analysis of Histone Modifications During Decidualization

We describe the genome-wide changes in histone modifications in human endometrial stromal cells during decidualization using chromatin immunoprecipitation combined with next-generation sequencing (ChIP sequence), which was recently reported by our lab [24].

Table 8.1 The number of the regions or genes with altered signals of histone modifications by decidualization [24]

Histone modifications	Regions with increased histone modification signals (gene)	Regions with decreased histone modification signals (gene)
H3K27ac	3705 (1846)	42 (39)
H3K4me3	945 (847)	109 (105)
H3K4me1	3 (3)	6 (7)
H3K27me3	2 (2)	5 (6)

Human endometrial stromal cells (ESC) were incubated with or without estradiol (10^{-8} M) and medroxyprogesterone acetate (10^{-6} M) for 14 days to induce decidualization. Histone modification signals [H3K27ac, H3K4me3, H3K4me1 (active marks), and H3K27me3 (repressive mark)] were searched every 1 kb genomic region between −10 kb from TSS and +10 kb from TTS. The altered (increased or decreased) signals by decidualization were analyzed by the difference in the signals between non-decidualized ESC and decidualized ESC. Regions showing more than a twofold increase or decrease in signals between non-decidualized ESC and decidualized ESC were defined as having increased or decreased histone modifications, respectively. The common regions of the two individuals were considered as the regions in which histone modification signals were altered by decidualization. The number of these regions and their adjacent genes are shown

8.4.1 Histone Modifications Altered by Decidualization

Induction of decidualization altered the H3K27ac and H3K4me3 statuses at many genomic regions, and that most of the histone modification changes were the increased signals (Table 8.1). On the other hand, decidualization altered the H3K4me1 and H3K27me3 statuses at only a few regions. However, decidualization was previously reported to cause many H3K27me3 changes [11]. The discrepancy may be due to the difference in the method, as the previous study used a ChIP-on-chip assay and induced decidualization with cAMP and medroxyprogesterone acetate (MPA), while our study used a ChIP sequence and induced decidualization with estradiol (E) and MPA. In fact, the genes and signaling pathways regulated by E + MPA differ from those regulated by cAMP (in our unpublished data).

8.4.2 Genomic Location of the Altered Histone Modifications

The locations of the H3K27ac- and H3K4me3-increased regions were classified into three groups: upstream [−10 kb upstream from transcription start site (TSS)], gene body [between TSS and transcription termination site (TTS)] including introns and exons, and intergenic (+10 kb downstream from TTS). Most of the H3K27ac- or H3K4me3-increased regions were located in the gene body (H3K27ac: 73.5 %, H3K4me3: 60.1 %).

The location of these regions was further analyzed based on the distance from the TSS of their adjacent genes. The location of the H3K27ac- or H3K4me3-increased regions was classified into three groups: proximal promoter (within ± 3 kb region from TSS), distal upstream promoter (more than 3 kb upstream from

TSS), and distal downstream promoter (more than 3 kb downstream from TSS). The H3K27ac-increased regions were predominantly located in the distal promoter regions (distal upstream and distal downstream) (79.4 %), and half of the H3K4me3-increased regions (50.1 %) were also located in the distal promoter regions. These findings indicate that histone modifications occur in the regions distant from the proximal region around the TSS in human endometrial stromal cells undergoing decidualization. This is consistent with previous reports that H3K27ac changes occur in the distal promoter regions as well as in the proximal promoter region [16, 25, 26].

8.4.3 Relationship Between H3K27ac or H3K4me3 and Gene Expression

Genome-wide mRNA levels were analyzed by RNA sequence. Decidualization upregulated 881 genes and downregulated 1379 genes (>1.4-fold changes). Of upregulated 881 genes, 223 (25.3 %) had either or both of the H3K27ac- and H3K4me3-increased regions (H3K4me3, 2.2 %; H3K27ac, 16.9 %; both H3K4me3 and H3K27ac, 6.2 %), raising the possibility that they are regulated by the increases of H3K27ac and H3K4me3 (Fig. 8.1). On the other hand, most (97.7 %) of the downregulated 1379 genes by decidualization did not have the H3K27ac- or H3K4me3-decreased regions.

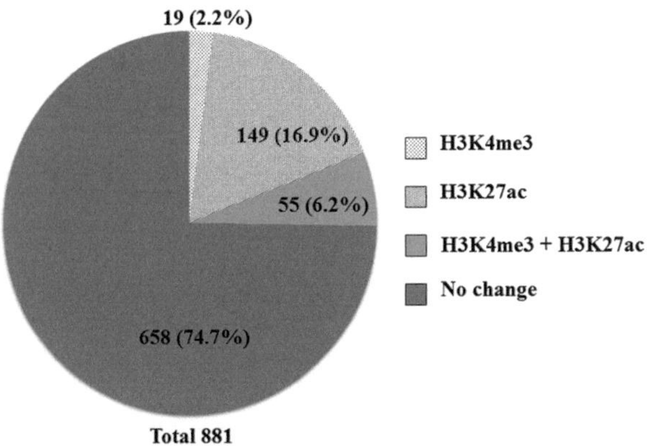

Fig. 8.1 Number of genes with the increased histone modifications in genes upregulated by decidualization [24]. Genome-wide mRNA levels were analyzed by RNA sequence in decidualized-human endometrial stromal cells (ESC) and non-decidualized ESC. A total of 881 genes were commonly upregulated by decidualization stimuli in two individuals. The upregulated genes were classified into four groups based on whether they have H3K27ac- or H3K4me3-increased regions: genes without H3K27ac and H3K4me3 (658 genes), genes with H3K4me3 (19 genes), genes with H3K27ac (149 genes), and genes with both H3K27ac and H3K4me3 (55 genes)

Regarding the association of the increase of H3K27ac or H3K4me3 with the increase in mRNA levels, decidualization-induced increases in mRNA levels were greater in regions with increases in H3K27ac- or H3K4me3 than in other regions, which suggests that the increases of these histone modifications are involved in the upregulation of the gene expression during decidualization.

In the 223 upregulated genes with the increase in H3K27ac or H3K4me3 by decidualization shown in Fig. 8.1, 43 had H3K27ac- or H3K4me3-increased regions in the proximal promoter region, 108 in the distal promoter regions, and 72 in both proximal and distal promoter regions (data not shown). Decidualization-induced increases in mRNA levels were greater whenever the increase in H3K27ac or H3K4me3 is present in either the proximal or distal promoter region. The presence of H3K27ac- or H3K4me3-increased regions in both proximal and distal promoter regions showed the synergistic effect on the increase in mRNA levels by decidualization. These results suggest that the increase of H3K27ac or H3K4me3 even in the distal promoter region is involved in the upregulation of the gene expression during decidualization. In fact, the presence of H3K27ac- or H3K4me3-increased regions in the distal promoter regions as well as in the proximal promoter region has been reported to activate gene expression [16, 25, 26]. Changes of H3K27ac and H3K4me3 statuses in the distal promoter region affect gene expression levels during cell differentiation [27, 28]. Recently, not only the regions near TSS but also the distal promoter regions are considered as important regions for transcription. Long-range chromatin interactions, such as distal enhancer-proximal promoter interactions, are recognized as an important mechanism to regulate gene expression levels [29, 30]. Therefore, the result on human endometrial stromal cells also strongly suggests that the interaction between the distal and proximal promoter regions is involved in the upregulation of gene expression during decidualization (Fig. 8.2).

8.4.4 *Biological Roles of Increases in H3K27ac and H3K4me3*

The potential roles of increases of H3K27ac and H3K4me3 during decidualization was investigated using a KEGG pathway enrichment analysis in the upregulated 223 genes with the increase in H3K27ac or H3K4me3. Three pathways were identified as significantly enriched pathways: type II diabetes mellitus, insulin signaling pathway, and aldosterone-regulated sodium reabsorption (Table 8.2). They commonly included the insulin signaling-related genes: insulin receptor substrate 1 (IRS1), insulin receptor substrate 2 (IRS2), insulin receptor (INSR), forkhead box O1 (FOXO1), v-akt murine thymoma viral oncogene homolog 3 (AKT3), and mitogen-activated protein kinase 10 (MAPK10). Insulin signaling has roles in reproductive tissues including the ovary and endometrium [31, 32]. The expression levels of several insulin signaling-related genes such as IRS1, IRS2,

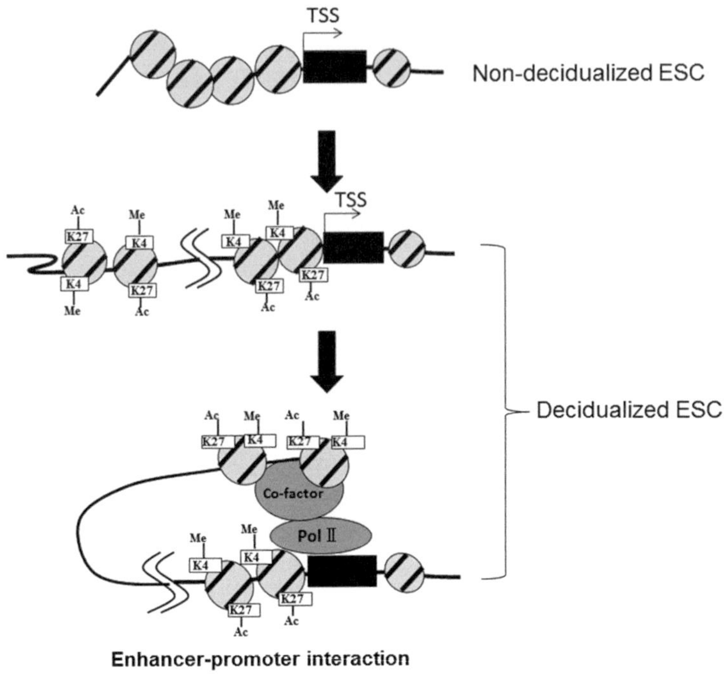

Fig. 8.2 Enhancer-promoter interactions in the regulation of gene expression. Regions considered important for transcription include both regions near the transcription start site (TSS) and distal promoter regions. Long-range chromatin interactions, such as distal enhancer-proximal promoter interactions, are known to regulate gene expression levels. Increases in H3K27ac and H3K4me3 are observed in the distal promoter region in human endometrial stromal cells during decidualization, suggesting that the interaction between the distal and proximal promoter regions is involved in the upregulation of gene expression during decidualization

INSR, and FOXO1 increase during decidualization in human endometrial stromal cells [33, 34]. Those genes are reported to be involved in glucose uptake in various types of cells [35]. In human endometrial stromal cells, glucose uptake has been found to be significantly increased by decidualization [24]. Interestingly, when human endometrial stromal cells were incubated in the presence of E + MPA, the increase in mRNA expression of prolactin (a specific marker of decidualization) was significantly lower at low glucose concentrations (0 and 5 mM) than at the normal glucose concentration (24 mM). This suggests that glucose is necessary for human endometrial stromal cells to undergo decidualization, which is also consistent with previous reports [36, 37]. Taken together, these findings strongly suggest that stimulating decidualization activates insulin signaling-related genes by increasing histone modifications of H3K27ac or H3K4me3, which in turn contribute to decidualization through the increase of glucose uptake.

Table 8.2 Enriched signaling pathways of upregulated genes with H3K27ac- or H3K4me3-increased regions by decidualization [24]

Pathway	Genes	Benjamini *P* value	*P* value
Type II diabetes mellitus	IRS1, IRS2, INSR, PIK3CG, MAPK10, HK2	0.040308	8.23E-04
Insulin signaling pathway	IRS1, IRS2, INSR, PIK3CG, MAPK10, FOXO1, AKT3, PRKAB2, HK2,	0.041218	0.001262
Aldosterone-regulated sodium reabsorption	IRS1, IRS2, INSR, PIK3CG, SGK1, HSD11B1	0.042383	4.33E-04

Three pathways were identified as significantly enriched pathways in the upregulated 223 genes with H3K27ac- or H3K4me3-increased regions by decidualization (KEGG pathway enrichment analysis)

8.5 DNA Methylation During Decidualization

Genome-wide change in DNA methylation was investigated using an Illumina Infinium HumanMethylation 450 BeadChip in human endometrial stromal cells treated with or without E2 and MPA for 14 days. This analysis covers DNA methylation in 480,000 CpG sites from 1500 bp upstream of TSS to 1000 bp downstream of TTS. When the difference of DNA methylation rate at CpG sites between the two groups was considered significant if it was more than 20 %, only 171 sites (0.0035 %) were differentially methylated. These included 58 hypomethylation and 113 hypermethylation sites (Table 8.3). This suggests that decidualization did not remarkably change the DNA methylation status in human endometrial stromal cells. Recently, Houshdaran et al. [13] compared DNA methylation status of the endometrial tissues between proliferative and mid-secretory phases using an Illumina Infinium HumanMethylation 27 BeadChip that covers DNA methylation in 27,000 CpG sites of the promoter region around the TSS. In that study, the cut-off value of the difference of DNA methylation rate between the two groups was set at 13.5 %, and only 67 differentially methylated CpG sites were identified. These results suggest that DNA methylation in the endometrial tissues does not remarkably change between the proliferative and mid-secretory phases. Taken together, these results indicate that it is unlikely that DNA methylation is involved in the regulation of gene expression in human endometrial stromal cells during decidualization.

8.6 miRNA Profiles During Decidualization

A previous study using a microarray found that expression of miRNAs in human decidualized stromal cells (i.e., cells treated with cAMP and MPA for 4 days) was different from that in non-decidualized cells [38]. In that study, a total of 16 miRNAs were upregulated and 33 miRNAs downregulated. A miRNA target

Table 8.3 Differentially methylated CpG sites between non-decidualized and decidualized cells

	The number of differentially methylated probe		
Difference of beta value	Total	Hypomethylated	Hypermethylated
0.20<	171 (0.0035 %)	58	113
0.30<	23 (0.00048 %)	8	15

Human endometrial stromal cells were treated with or without E and MPA for 14 days. DNA methylation was analyzed by Illumina Infinium HumanMethylation 450 BeadChip. When the difference of DNA methylation rate (beta value) between the two groups was considered significant if it was more than 0.2 (20 %) and 0.3 (30 %), differentially methylated CpG sites were 171 (0.0035 %) consisting of 58 hypomethylation and 113 hypermethylation and 23 (0.00048 %) consisting of 8 hypomethylation and 15 hypermethylation, respectively

prediction algorithm predicted that the target genes were transcription factors (FOXO1, HOXOA10, HMGA2, STAT5, CREBBP), extracellular matrix remodeling enzymes (MMP, FN), interleukin families, cell cycle regulators (CDK, CDKN, MYC, PTEN), and growth factors (IGF, VEGFA, VEGFB, BDNF). A recent study also reported that 26 miRNAs were upregulated and 17 miRNAs were downregulated in decidualized cells [induced with progesterone (10^{-6} M) and E (3×10^{-8} M) for 9 days] compared with non-decidualized cells [39]. The potential target genes were transcription factors (FOXO1, CEBPB, CREB, SP-1, HOXOA10, STAT5), extracellular matrix remodeling enzymes (TIMP), interleukin families, and growth factors and their receptors (IGF, IGFR, VEGFA, EGFR, TGFB, TGFBR). These miRNAs may affect differentiation and decidualization events by regulating expression of their target genes. However, only two miRNAs (miR-181b and miR-181d) were commonly identified in the two studies. Further studies are needed to better understand the functional role of miRNAs in the regulation of gene expression in human endometrial stromal cells during decidualization.

8.7 Conclusions

In this chapter, we focused on the involvement of epigenetic mechanisms such as histone modifications, DNA methylation, and miRNAs in the upregulation of gene expression in human endometrial stromal cells during decidualization (Fig. 8.3). During decidualization, human endometrial stromal cells undergo genome-wide changes in histone modification statuses. The main changes were increases of H3K27ac and H3K4me3, which are active marks, in a number of genomic regions. The increases of H3K27ac or H3K4me3 occurred in both the proximal and distal promoter regions, suggesting that interactions between the distal enhancer and proximal promoter regions are involved in the upregulation of gene expression in human endometrial stromal cells during decidualization. Interestingly, the increases of H3K27ac and H3K4me3 during decidualization were associated with activation

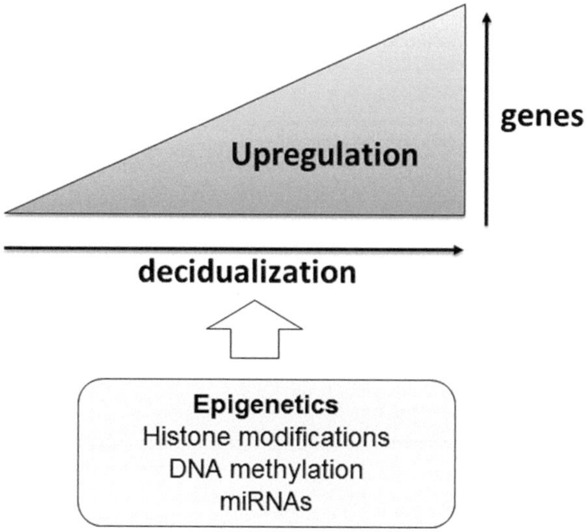

Fig. 8.3 Schematic showing involvement of epigenetics in the upregulation of gene expression in human endometrial stromal cells during decidualization

of the insulin signaling pathway, which in turn contributes to decidualization through the increase of glucose uptake.

Decidualization is a naturally occurring differentiation process in the human endometrium during the menstrual cycle and is accompanied by marked functional changes. Gene regulation during the process of cell differentiation is closely associated with the dynamic changes of histone modifications. Most of the global chromatin studies so far have focused on histone modification statuses associated with self-renewal of stem cells or their differentiation into mature cell types. It is interesting to note that genome-wide changes in histone modification statuses occur even in well differentiated somatic cells, which are human endometrial stromal cells.

References

1. Salker M, Teklenburg G, Molokhia M, Lavery S, Trew G, Aojanepong T, Mardon HJ, et al. Natural selection of human embryos: impaired decidualization of endometrium disables embryo-maternal interactions and causes recurrent pregnancy loss. PLoS ONE. 2010;5, e10287.
2. Laird SM, Tuckerman EM, Li TC. Cytokine expression in the endometrium of women with implantation failure and recurrent miscarriage. Reprod Biomed Online. 2006;13:13–23.
3. Wang W, Taylor RN, Bagchi IC, Bagchi MK. Regulation of human endometrial stromal proliferation and differentiation by C/EBPbeta involves cyclin E-cdk2 and STAT3. Mol Endocrinol. 2012;26:2016–30.
4. Aghajanova L, Horcajadas JA, Weeks JL, Esteban FJ, Nezhat CN, Conti M, et al. The protein kinase A pathway-regulated transcriptome of endometrial stromal fibroblasts reveals compromised differentiation and persistent proliferative potential in endometriosis. Endocrinology. 2010;151:1341–55.

5. Popovici RM, Kao LC, Giudice LC. Discovery of new inducible genes in in vitro decidualized human endometrial stromal cells using microarray technology. Endocrinology. 2000;141:3510–3.
6. Matsuoka A, Kizuka F, Lee L, Tamura I, Taniguchi K, Asada H, et al. Progesterone increases manganese superoxide dismutase expression via a cAMP-dependent signaling mediated by non-canonical Wnt5a pathway in human endometrial stromal cells. J Clin Endocrinol Metab. 2010;95:E291–9.
7. Li B, Carey M, Workman JL. The role of chromatin during transcription. Cell. 2007;128:707–19.
8. Tamura I, Taketani T, Lee L, Kizuka F, Taniguchi K, Maekawa R, et al. Differential effects of progesterone on COX-2 and Mn-SOD expressions are associated with histone acetylation status of the promoter region in human endometrial stromal cells. J Clin Endocrinol Metab. 2011;96:E1073–82.
9. Krusche CA, Vloet AJ, Classen-Linke I, von Rango U, Beier HM, Arlfer J. Class I histone deacetylase expression in the human cyclic endometrium and endometrial adenocarcinomas. Hum Reprod. 2007;22:2956–66.
10. Sakai N, Maruyama T, Sakurai R, Masuda H, Yamamoto Y, Shimizu A, et al. Involvement of histone acetylation in ovarian steroid-induced decidualization of human endometrial stromal cells. J Biol Chem. 2003;278:16675–82.
11. Grimaldi G, Christian M, Steel JH, Henriet P, Poutanen M, Brosens JJ. Down-regulation of the histone methyltransferase EZH2 contributes to the epigenetic programming of decidualizing human endometrial stromal cells. Mol Endocrinol. 2011;25:1892–903.
12. Yamagata Y, Asada H, Tamura I, Lee L, Maekawa R, Taniguchi K, et al. DNA methyltransferase expression in the human endometrium: down-regulation by progesterone and oestrogen. Hum Reprod. 2009;24:1126–32.
13. Houshdaran S, Zelenko Z, Irwin JC, Giudice LC. Human endometrial DNA methylation is cycle-dependent and is associated with gene expression regulation. Mol Endocrinol. 2014;28:1118–35.
14. Riggs AD, Martienssen RA, Russo VEA. Introduction. In: Russo VEA, Martienssen RA, Riggs AD, editors. Epigenetic mechanisms of gene regulation. New York: Cold Spring Harbor Laboratory Press; 1996. p. 1–4.
15. Fischle W, Wang Y, Allis CD. Binary switches and modification cassettes in histone biology and beyond. Nature. 2003;425:475–9.
16. Creyghton MP, Cheng AW, Welstead GG, Kooistra T, Carey BW, Steine EJ, et al. Histone H3K27ac separates active from poised enhancers and predicts developmental state. Proc Natl Acad Sci U S A. 2010;107:21931–6.
17. Barski A, Cuddapah S, Cui K, Roh TY, Schones DE, Wang Z, et al. High-resolution profiling of histone methylations in the human genome. Cell. 2007;129:823–37.
18. Zhang JA, Mortazavi A, Williams BA, Wold BJ, Rothenberg EV. Dynamic transformations of genome-wide epigenetic marking and transcriptional control establish T cell identity. Cell. 2012;149:467–82.
19. Ballestar E, Wolffe AP. Methyl-CpG-binding proteins. Targeting specific gene repression. Eur J Biochem. 2001;268:1–6.
20. Yagi S, Hirabayashi K, Sato S, Li W, Takahashi Y, Hirakawa T, et al. DNA methylation profile of tissue-dependent and differentially methylated regions (T-DMRs) in mouse promoter regions demonstrating tissue-specific gene expression. Genome Res. 2008;18:1969–78.
21. He L, Hannon GJ. MicroRNAs; small RNAs with a big role in gene regulation. Nat Rev Genet. 2004;5:522–31.
22. Tamura I, Asada H, Maekawa R, Tanabe M, Lee L, Taketani T, et al. Induction of IGFBP-1 expression by cAMP is associated with histone acetylation status of the promoter region in human endometrial stromal cells. Endocrinology. 2012;153:5612–21.
23. Tamura I, Sato S, Okada M, Tanabe M, Lee L, Maekawa R, et al. Importance of C/EBPβ binding and histone acetylation status in the promoter regions for induction of IGFBP-1,

PRL and Mn-SOD by cAMP in human endometrial stromal cells. Endocrinology. 2014;155:275–86.
24. Tamura I, Ohkawa Y, Sato T, Suyama M, Jozaki K, Okada M, et al. Genome-wide analysis of histone modifications in human endometrial stromal cells. Mol Endocrinol. 2014;28:1656–69.
25. Pham TH, Benner C, Lichtinger M, Schwarzfischer L, Hu Y, Andreesen R, et al. Dynamic epigenetic enhancer signatures reveal key transcription factors associated with monocytic differentiation states. Blood. 2012;119:e161–71.
26. Mikkelsen TS, Xu Z, Zhang X, Wang L, Gimble JM, Lander ES, et al. Comparative epigenomic analysis of murine and human adipogenesis. Cell. 2010;143:156–69.
27. Tian Y, Jia Z, Wang J, Huang Z, Tang J, Zheng Y, et al. Global mapping of H3K4me1 and H3K4me3 reveals the chromatin state-based cell type-specific gene regulation in human Treg cells. PLoS ONE. 2011;6, e27770.
28. Cui K, Zang C, Roh TY, Schones DE, Childs RW, Peng W, et al. Chromatin signatures in multipotent human hematopoietic stem cells indicate the fate of bivalent genes during differentiation. Cell Stem Cell. 2009;4:80–93.
29. Li G, Ruan X, Auerbach RK, Sandhu KS, Zheng M, Wang P, et al. Extensive promoter-centered chromatin interactions provide a topological basis for transcription regulation. Cell. 2012;148:84–98.
30. Zhang Y, Wong CH, Birnbaum RY, Li G, Favaro R, Ngan CY, et al. Chromatin connectivity maps reveal dynamic promoter-enhancer long-range associations. Nature. 2013;504:306–10.
31. Purcell SH, Chi MM, Lanzendorf S, Moley KH. Insulin-stimulated glucose uptake occurs in specialized cells within the cumulus oocyte complex. Endocrinology. 2012;153:2444–54.
32. Lathi RB, Hess AP, Tulac S, Nayak NR, Conti M, Giudice LC. Dose-dependent insulin regulation of insulin-like growth factor binding protein-1 in human endometrial stromal cells is mediated by distinct signaling pathways. J Clin Endocrinol Metab. 2005;90:1599–606.
33. Ganeff C, Chatel G, Munaut C, Frankenne F, Foidart JM, Winkler R. The IGF system in in-vitro human decidualization. Mol Hum Reprod. 2009;15:27–38.
34. Kim JJ, Buzzio OL, Li S, Lu Z. Role of FOXO1A in the regulation of insulin-like growth factor-binding protein-1 in human endometrial cells: interaction with progesterone receptor. Biol Reprod. 2005;73:833–9.
35. Saltiel AR, Kahn CR. Insulin signalling and the regulation of glucose and lipid metabolism. Nature. 2001;414:799–806.
36. Frolova AI, Moley KH. Quantitative analysis of glucose transporter mRNAs in endometrial stromal cells reveals critical role of GLUT1 in uterine receptivity. Endocrinology. 2011;152:2123–8.
37. Frolova A, Flessner L, Chi M, Kim ST, Foyouzi-Yousefi N, Moley KH. Facilitative glucose transporter type 1 is differentially regulated by progesterone and estrogen in murine and human endometrial stromal cells. Endocrinology. 2009;150:1512–20.
38. Qian K, Hu L, Chen C, Li H, Liu N, Li Y, et al. Has-miR-222 is involved in differentiation of endometrial stromal cells in vitro. Endocrinology. 2009;150:4734–43.
39. Estella C, Herrer I, Moreno-Moya JM, Quinonero A, Martinez S, Pellicer A, et al. miRNA signature and dicer requirement during human endometrial stromal decidualization in vitro. PLoS ONE. 2012;7, e41080.

Chapter 9
Stem/Progenitor Cells in the Human Endometrium

Tetsuo Maruyama

Abstract Adult stem cells, also termed tissue stem cells or somatic stem cells, are rare populations residing in almost all adult tissues. They can self-renew and possess the capacity for multi-lineage differentiation. They play critical roles in tissue homeostasis, regeneration, repair, and response to injury. It has long been believed that stem/progenitor cells exist in the human endometrium based on its unique capacity to regenerate and regress cyclically in response to fluctuating ovarian steroid hormones during each menstrual cycle throughout a reproductive life. There is increasing evidence that human endometrium contains small populations of epithelial progenitor cells (EPCs), mesenchymal stem cells (MSCs), and side population cells (SPCs) that are likely responsible for its monthly regeneration and tissue homeostasis. This review summarizes the identification of EPCs, MSCs, and SPCs and discusses how they are involved in the physiological remodeling and regeneration of the human endometrium.

Keywords Endometrium • Progenitor cells • Mesenchymal stem cells • Side population • Endometrial regeneration

9.1 Tissue Stem/Progenitor Cells

Stem cells are defined as undifferentiated cells with a capacity for both self-renewal and the ability to differentiate into one or more lineages of more mature and specialized cells [1]. Stem cells include embryonic stem cells (ESCs), induced pluripotent stem cells (iPSCs), and adult/somatic stem cells (SSCs). Germline stem cells (GSCs) are categorized as SSCs when they are derived from tissues and are capable of giving rise to haploid gametes, sperm, or oocytes. SSCs are multipotent but not pluripotent. That is, they are able to generate multiple types of differentiated cells that are generally limited to those of the original tissue, organ, or physiological system to which the SSCs belong [1].

T. Maruyama (✉)
Department of Obstetrics and Gynecology, Keio University School of Medicine,
35 Shinanomachi, Shinjuku-ku, Tokyo 160-8582, Japan
e-mail: tetsuo@keio.jp

H. Kanzaki (ed.), *Uterine Endometrial Function*,
DOI 10.1007/978-4-431-55972-6_9

Hematopoietic stem cells were the first type of SSCs to be isolated and utilized therapeutically through bone marrow transplantation [2]. Subsequently, the existence, physiological role, and therapeutic potentials of various other types of SSCs including mesenchymal stem cells (MSCs) have been demonstrated. In particular, MSCs have been extensively studied, and they are considered one of the most promising biological materials for regenerative medicine [3]. MSCs are plastic-adherent clonogenic cells with a characteristic surface phenotype. They are capable of differentiating in vitro into multiple mesodermal lineages, including adipocytes, osteoblasts, and chondrocytes [4]. It is generally accepted that SSCs in the blood, skin, and gut support tissue homeostasis and regeneration, whereas SSCs in other tissues such as muscle, liver, kidney, and lung play pivotal roles in tissue repair and responses to injury [5].

A stem cell is thought to produce a progenitor cell through asymmetric cell division. In contrast to the multipotency of a stem cell, a progenitor cell is monopotential or at most oligopotential in that it is more committed than a stem cell to differentiate into its "target" cell(s) [6]. More importantly, stem cells are believed to replicate indefinitely, whereas progenitor cells are able to divide a limited number of times [6]. Whereas "progenitor cells" and "stem cells" differ quantitatively and qualitatively, progenitors play similar important roles in tissue homeostasis, regeneration, repair, and response to injury. In this review "progenitor cell" and "stem cell" are generally equated.

9.2 Endometrial Stem/Progenitor Cells

It is believed that endometrial stem/progenitor cells mediate endometrial regeneration [7–9]. This hypothesis is supported by clinical observations, studies of cell proliferation, and analyses of gland monoclonality [10, 11]. Loss of the endometrial functionalis layer at menstruation and its subsequent regeneration from the endometrial basalis suggest that endometrial stem/progenitor cells are probably present in the basalis [7–9]. Those cells may give rise to more differentiated endometrial cell components responsible for the regeneration of the functionalis layer of the endometrium [12]. Several research groups, including ours, have employed a variety of methods to identify, isolate, and/or characterize putative endometrial stem/progenitor cells capable of multipotent differentiation. [13–16].

9.2.1 Human Endometrial Epithelial Progenitor Cells

In vitro clonogenicity at extremely low seeding densities is a characteristic of stem/progenitor cells and has been demonstrated in many types of adult stem cells [10, 17]. Putative endometrial epithelial progenitor cells were first identified as clonogenic cells in cycling human endometrium [18]. Epithelial colony-forming units (CFU) consist of two types: large (0.8 %) and small (0.14 %). Large epithelial CFU could be serially cloned at least three to four times at very low seeding densities (10–20 cell/cm^2), substantiating a high self-renewal activity in vitro, whereas small CFU lacked that ability [19].

Recently, SSEA-1 (CD15) was demonstrated to be a surface marker that was differentially expressed by epithelial cells of the basalis and those of the functionalis [20]. Furthermore, SSEA-1+ epithelial cells generated spheroids with an epithelial polarity in 3D Matrigel cultures and exhibited longer telomeres and greater telomerase activity, indicating that they have some adult stem/progenitor cell-like properties [20]. However, whether the SSEA-1+ population includes clonogenic and self-renewing epithelial cells remains unknown.

9.2.2 Endometrial Mesenchymal Stem/Stromal Cells

Putative endometrial stromal stem/progenitor cells were first identified as a rare clonogenic cell population (1.25 %) in single cell suspensions derived from human endometrium [18]. Like epithelial CFU, stromal CFU were divided into two types: large (0.02 %) and small (1.23 %). Large stromal CFU could be serially cloned at least three to four times when seeded at very low densities (10–20 cell/cm^2), indicating substantial self-renewal capacity in vitro. However, this property was not observed in small stromal CFU. In addition, large CFU were positive for typical MSC markers and had the potential for multi-lineage differentiation into adipocytes, osteoblasts, myocytes, and chondrocytes, suggesting MSC properties [19].

The combined use of CD140b (platelet-derived growth factor receptor β, PDGFRB) and CD146 antibodies or the single use of SUSD2 (W5C5) antibody has been employed to prospectively isolate human endometrial MSCs (eMSCs) [21–23]. Endometrial cells positive for these markers showed a perivascular location close to endothelial cells in the endometrium [13, 14], raising the possibility that human eMSCs may be identical to pericytes [21, 23]. Indeed, gene expression profiling of human eMSCs revealed the transcription of genes involved in angiogenesis, inflammation, and immunomodulation. Furthermore, gene expression data substantiated their pericyte identity as well as responsiveness to steroid hormones, hypoxic conditions, and capacities for cell communication, proteolysis, self-renewal, and multipotency [24]. Furthermore, human SUSD2+ eMSC generated human vimentin+ endometrial stroma-like tissue and migratory endothelial cells in the kidney parenchyma when transplanted under the kidney capsule of severely immunodeficient mice [23].

Taken together these findings suggest that eMSCs are clonogenic and likely multipotent pericytes. It appears that they can respond to hormonal changes and thereby contribute to the regeneration of endometrial stroma and vasculature during the menstrual cycle [13, 14]. It has been demonstrated that eMSCs are involved in endometrial stromal decidualization to support embryo implantation and early placentation through secretion of immunomodulatory factors [25].

Very recently, it was reported that expression of an epigenetic stemness-related signature (including high mobility group protein 2 [HMGB2]) was lost in human endometrial stromal cells (HESCs) collected from patients with recurrent pregnancy loss (RPL) [26]. Further analysis revealed that endometrial clonogenic cell populations are significantly reduced in RPL patients compared to the control group. Given that the loss of HMGB2 perturbs decidualization and promotes senescence in HESCs, endometrial stem cell deficiency may cause RPL through the triggering of accelerated stromal senescence and impaired decidualization [26].

9.2.3 Human Endometrial Side Population Cells

Primitive, undifferentiated cells in many species can be identified by treatment with Hoechst 33342 dye followed by flow cytometric analysis. The primitive fraction has been termed the "side population" (SP) due to its low fluorescence and consequent offset position in flow cytometric dot plots. The low fluorescence is due to the continuous removal of Hoechst dye via a cell membrane transporter, adenosine triphosphate (ATP)-binding cassette subfamily G member 2 (ABCG2) [29]. Adult stem cells in many different types of tissues, organs, and species are found in the SP fraction [27, 28].

To isolate SP cells from human endometrium, we dissociated human endometria mechanically and enzymatically, obtained epithelial-enriched and stromal-enriched fractions, stained both fractions with Hoechst dye, and subjected them to flow cytometric analysis and cell sorting (Fig. 9.1a, b) [30]. We found that each preparation contained a small subset of cells in the SP fraction (Fig. 9.1b). The appearance of the SP populations was blocked by 50 μM reserpine, an inhibitor of ABCG2 (Fig. 9.1b, insets), a general characteristic of SP cells [27, 28]. Because it was unclear which SP fraction contained the endometrial stem/progenitor cells, we mixed the epithelial and stromal SP cells or isolated SP cells from a mixture of the two fractions. We designated the SP and main population (MP) cells derived from the mixture as endometrial SP (ESP) and endometrial MP (EMP). The endometrial replicative population was designated as the endometrial replicative population (ERP) (Fig. 9.1b). The ESP and EMP cells were then subjected to in vitro and in vivo experiments for further characterization and functional analysis [30].

Immunofluorescence studies of ESP cultures revealed that there existed three distinct types of colonies consisting of either fibroblastoid stromal cells positive for CD13, endothelial progenitor-like cells positive for CD31, or epithelial cells positive for cytokeratin (CK) (Fig. 9.1c). In contrast to ESP cultures, most of the colonies formed in the EMP cultures consisted of CD13-positive fibroblastic stromal cells (Fig. 9.3c).

To elucidate the stem cell-like regenerative capacity of ESP cells in vivo, we transplanted ESP cells under the kidney capsule of ovariectomized NOD/SCID/γc

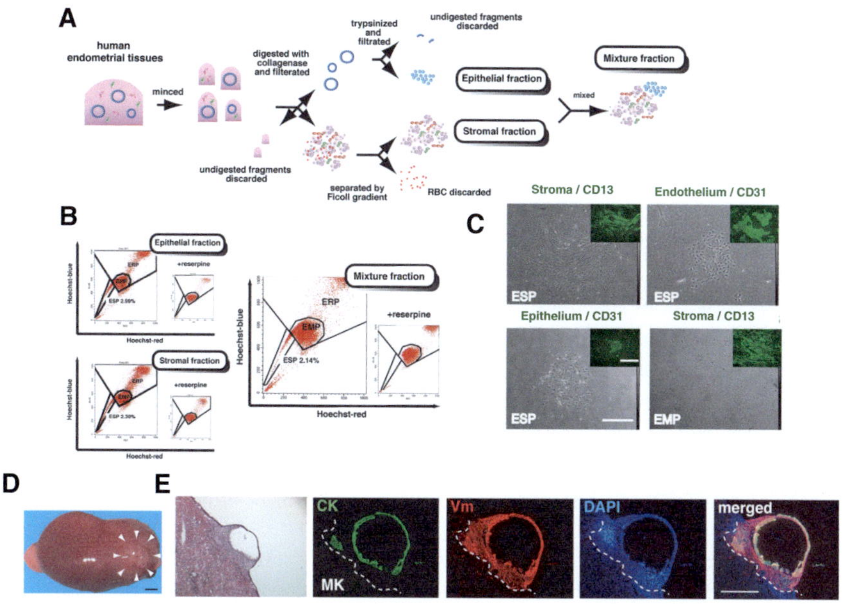

Fig. 9.1 Isolation, identification, and functional analysis of SP cells derived from human endometrium (Adapted from Masuda et al. [30]). (**a**) Summary of procedures for the preparation of epithelial-enriched and stromal-enriched fractions and a mixture of both fractions from human cycling endometria. RBC, red blood cell. (**b**) Flow cytometric distribution of ESP cells, EMP cells, and the endometrial replicative population (ERP) in each of the three fractions stained with Hoechst 33342. Addition of 50 μM reserpine resulted in the disappearance of the ESP fraction (*inset* in each panel). (**c**) Phase contrast micrographs and fluorescence images (*insets*) of colonies generated from ESP and EMP cultures. The ESP cells and EMP cells were separately seeded at a clonal density, cultured in EGM-2MV medium for 2 weeks, and subjected to immunofluorescence studies using antibodies against the indicated markers. CK, cytokeratin. Bars, 500 μm. (**d**) The macroscopic appearance of an ESP-initiated lesion (surrounded by *white arrowheads*) in the kidney of a NOD/SCID/γ_c^{null} (NOG) mouse treated with E_2 pellets for 10 weeks. Bar, 1 mm. (**e**) H&E staining and immunofluorescence images of section of the same lesion indicated in (**d**). Immunofluorescence images were co-stained with DAPI and antibodies against CK and vimentin (Vm). The borders between the reconstituted tissue and the mouse kidney (MK) are indicated by the *dotted lines*. Bars, 500 μm

null (NOG) mice followed by treatment with 17β-estradiol (E_2) for 8–10 weeks. ESP cells, but not EMP cells, generated a cystic mass (Fig. 9.1d) with delineated CK-positive glandular structures and vimentin-positive stromal structures at the site of transplantation (Fig. 9.1e). However, the reconstitution efficiency of the complete endometrium-like tissue was low in that the reconstitution was observed in only 2 out of 24 xenotransplanted mice (Fig. 9.1e). Importantly, despite the low reconstitution efficiency, ESP cells produced various endometrial cell components including endothelial cells, stromal cells, and smooth muscle cells in the remaining (22 of 24) xenotransplanted mice [30]. Thus, ESP cells gave rise to several lineages of endometrial cells and therefore possessed stem/progenitor-like cell characteristics.

ESP cells were also identified in short-term cultures of endometrial cells [30]. We designated ESP cells isolated from cultured endometrial cells as cESP [31–33] while ESP cells freshly isolated from endometrium were designated as fESP [30, 32, 34]. cESP were originally negative for both epithelial (CD9) and stromal (CD13) markers but were able to differentiate into CD9+E-cadherin+ gland-like organoids and CD13+ stromal clusters in long-term Matrigel cultures [31]. As mentioned above, fESP gave rise to epithelial, stromal, and endothelial cells in vitro without Matrigel [30]. Although fESP cells are negative for estrogen receptor-α (ERα) and progesterone receptor (PR) and positive for ERβ, cESP alone can respond to treatment with E_2 and progesterone (P_4) to decidualize in vitro [32]. In addition to the subtle differences, there are critical differences in CFU activity between cESP and fESP [31, 32]. cESP have higher CFU activity than non-SP or the main and replicative population (MRP) cells [31, 32] while fESP cells had lower CFU activity [30]. Differences in cell cycle status between fESP and cESP (G0 and G1/S/G2/M, respectively) account for this differential CFU activity [32]. The quiescence of fESP cells accords with adult stem cell properties [35]. It is also possible that in vitro CFU and differentiation assays produce a microenvironment different from stem cell niche and therefore inappropriate for maintaining adult stem cell properties, resulting in differences between laboratories. The differences in the character and behavior between cESP and fESP have been discussed in detail elsewhere [15].

9.2.4 Bone Marrow-Derived Cells

Bone marrow is a possible source of human endometrial, epithelial, and stromal stem/progenitor cells [36, 37]. Clinical studies have showed that chimerism exists in the endometrial glands and stroma in patients who underwent transplantation of human leukocyte antigen-mismatched or sex-mismatched bone marrow [36, 38]. Furthermore, a transgenic mouse model tracking CD45+ HSCs identified BM-derived endometrial epithelium in a small number of cells in a small proportion of the animals [37], supporting an HSC origin for some endometrial epithelial cells. Thus, bone marrow stem cells might play a role in endometrial regeneration, at least in part in terms of cellular turnover and inflammatory stimuli [36]. Nevertheless, in women who underwent bone marrow transplantation from a male donor, XY+ endometrial cells were identified in the non-ESP, but not in the ESP progenitor fraction [39]. Given that ESP cells are putative endometrial stem/progenitor cells involved in endometrial regeneration [30–32, 34], the absence of bone marrow-derived cells in the ESP fraction suggests that bone marrow-derived stem cells are likely to contribute less to endometrial regeneration, particularly during each menstrual cycle, than stem/progenitor cells present in the endometrium itself. Most glands are originated from exclusively either donor or recipient, though some chimerism is observed within individual glands. This suggests that not all glands are monoclonal, which is consistent with the methylation pattern of individual glands [40].

9.2.5 *Menstrual Blood-Derived Mesenchymal Stem/Stromal Cells*

Menstrual blood contains MSC-like cells capable of differentiating into ectodermal, mesodermal, and endodermal lineages [41]. Given the perivascular location of eMSC in the functionalis [13], it is conceivable that eMSC and stromal fibroblasts may contribute to the multi-lineage differentiation activity of MSC that are isolated and cultured as plastic-adherent cells from menstrual blood. Also, it is tempting to speculate that circulating peripheral blood-derived stem/progenitor cells [42] may participate in the menstrual blood MSC activity through their influx into the endometrium and uterine cavity during menstruation. It is, however, unlikely that the MSC activity of menstrual blood is attributable to epithelial progenitor cells, in particular SSEA-1+ epithelial cells [20], because they are located in the basalis that fails to shed in the menstrual blood [20].

9.3 Role of Endometrial Stem/Progenitor Cells in Endometrial Function

9.3.1 *Regeneration of Human Endometrium*

The human endometrium undergoes cyclic changes, including proliferation, differentiation, tissue breakdown, and shedding (menstruation), more than 400 times throughout a reproductive life [16]. Human endometrial tissue can be structurally and functionally divided into two major compartments. The upper two-thirds (functional layer) are shed during menstruation in response to P_4 withdrawal and regenerate in the subsequent cycle. The lower basal layer consists of the bases of the endometrial glands surrounded by dense stroma. It persists after menstruation and is thought to supply cells for regenerating the functional layer each month. The restructuring of the functional layer is prerequisite for the growth and development of the endometrium ready for implantation or for menstruation [16]. Based on these features, one or more endometrium-specific tissue stem cell systems likely play a critical role in the regeneration and remodeling properties of the endometrium [7, 8].

Human endometrial regeneration begins with the rapid repair of the surface epithelium concomitantly or immediately after endometrial tissue breakdown [43]. It has long been thought that epithelial repair occurs through epithelial cell migration from basal gland stumps over the denuded surface [44–48]. However, Garry et al. have objected to this conventional concept. Instead, based on detailed histological, immunohistochemical, and scanning electron microscopic analyses, they proposed that cellular differentiation from stromal cells rather than direct extension from the residual basal epithelial glands contributes to endometrial surface epithelial regeneration [49, 50]. Alternatively, a gene profiling study raised the possibility that new surface epithelium might engulf remnants of the shedding functional layer during menstruation [43, 51]. In any case, once the luminal epithelium has been repaired, endometrial glands, stroma, and vasculature rapidly regenerate during the proliferative phase [13].

Endometrium is also able to regenerate postpartum, after operative removal, and in postmenopausal women undergoing estrogen replacement therapy. Although mice do not menstruate, they exhibit endometrial regeneration after parturition like humans. Cell fate mapping analyses using transgenic mice showed that endometrial epithelial tissue regeneration following parturition is at least partly attributable to the stromal mesenchymal-to-epithelial transition (MET) [52, 53]. The processes seem analogous to those observed in humans in which small stromal cells differentiate into or generate epithelial cells during reepithelialization after menstruation [49, 50].

9.3.2 Role of Putative Endometrial Stem/Progenitor Cells in the Regeneration of Endometrium

It has been postulated that endometrial stem/progenitor cells reside in the endometrial basalis. Endometrial progenitor cells give rise to a pool of rapidly proliferating transit amplifying epithelial, stromal, and vascular cells that regenerate new glands, stroma, and vasculature of the endometrial functionalis during the proliferative phase immediately after the luminal epithelium has resurfaced [13]. Furthermore, the postpartum, postoperative, and postmenopausal regeneration potential of human endometrium suggests that stem/progenitor cell populations reside deep in the basalis layer of the endometrium [13].

Despite the expectation that endometrial stem/progenitor cells are likely to be located at least predominantly in the basalis [7, 8], these candidate cells (including eMSC and ESP cells) are evenly divided between the basalis and the functionalis [13]. Human CD146+ and CD140b+ eMSCs are located perivascularly in the functionalis and basalis layers of the endometrium [21]. Perivascular SUSD2+ eMSC are also found in the functionalis and basalis layers [23]. ABCG2 is highly expressed in ESP cells [30], and therefore, endometrial ABCG2+ cells largely correspond to ESP cells, in particular fESP. Like eMSCs, endometrial ABCG2+ cells are located in and near the vascular wall of endometrial vessels in both functional and basal layers [30]. Intriguingly, most ABCG2-positive cells co-expressed CD31 and were preferentially located in small capillaries rather than large vessels [30].

The percentage of endometrial SP cells changes with the menstrual cycles, being highest in the early proliferative phase, decreasing in the early secretory phase, and lowest in the late secretory phase [30–32]. Given that the location of ESP cells in both the functionalis and basalis [30], it is likely that transit-amplifying cells and differentiated cells increase more profoundly than ESP cells from the proliferative to the secretory phase, resulting in the relative decrease in the percentage of ESPs.

Although differentiation is an important stem cell property, the SP is heterogeneous, presumably containing stem/progenitor cells of each endometrial cell lineage. Indeed, the endometrial SP contained CD326+ (EpCAM+) cells (epithelial, 27 %), CD10+ cells (stromal, 14 %), CD31+ cells (endothelial, 51 %), CD34+ cells (endothelial, hematopoietic cells, 46 %), and CD146+ cells (endothelial, MSC, 25 %) [54]. Of these lineages, CD31+, CD34+, CD146+, and CD140b+CD146+ cells (eMSCs) were significantly more abundant in endometrial SP cells than in MRP cells [54], substantiating the endothelial and MSC-like potential of endometrial SP cells. In addition to the overlap of surface markers, the similar perivascular location of eMSCs and some endometrial SP cells suggests that ESPs and eMSCs may be closely related as constituents of an as-yet-unidentified endometrial stem cell hierarchy [15].

As mentioned above, the regeneration of endometrial surface epithelium is presumably achieved through cellular differentiation from stromal cells and not proliferation from basal epithelial glands [49, 50]. ESP cells have been shown to generate endometrial epithelial and stromal cells in vitro and in vivo [30, 31]. It is possible that ESP cells alone or menstruating endometrial fragments containing ESP cells might be trapped within the uterine cavity after menstruation [55]. Such cells might contribute to endometrial regeneration. It is also possible that endometrial stem/progenitor daughter cells undergo a mesenchymal-to-epithelial transition (MET) or EMT and thereby contribute to endometrial regeneration during the human menstrual cycle [52, 53].

The perivascular localization of putative endometrial stem/progenitor cells in the functionalis suggests that they may be shed in the menstrual blood. Given the migratory, angiogenic, and/or endometrial tissue regeneration potentials of fESP and eMSC [13–15, 55], they may implant onto the peritoneum and other ectopic sites through retrograde menstruation and may behave as endometriosis stem/progenitor cells and/or endometriosis-initiating cells [55]. The possible role of endometrial stem/progenitor cells in the pathogenesis of endometriosis has been reviewed elsewhere [56–59].

9.3.3 *Experimental Model for Endometrial Regeneration and Angiogenesis*

Differences between humans and rodents make it difficult to study the regenerative and angiogenic processes in the endometrium. Thus, rodent data might not apply to human physiology. In vitro approaches for the modeling of physiologic processes may not accurately reflect and reproduce the normal processes of tissue shedding and regeneration.

To overcome these limitations, we developed a novel mouse model for human endometrial regeneration and regression using NOG mice in which human endometrial xenografts responded to cyclical exogenous sex hormones producing tortuous glands and differentiated decidual cells followed by large blood-filled cysts suggestive of menstruation [60]. We first prepared singly dispersed endometrial cells (SDECs) from human endometrium that included epithelial, stromal, endothelial, and immune cells. A limited number of SDECs was transplanted beneath the kidney capsule of NOG mice (Fig. 9.2a). A large cystic mass with fine surface vascular formation was generated at the transplanted site after 10 weeks of treatment with E_2 and P_4 (E_2+P_4) (Fig. 9.2b, upper panel and inset). H&E staining revealed that the mass consisted of endometrium-like tissue with well-delineated glands and stroma (Fig. 9.2b, lower panel). Moreover, subsequent P_4 withdrawal induced processes analogous to menstruation, including tissue breakdown, shedding, and bleeding, in the reconstituted secretory-phase endometrium-like tissue (Fig. 9.2c) [60].

Bioluminescent imaging (BLI) has been widely used to track the behavior of tumor, hematopoietic, and neural cells in living animals [61]. The novel mouse model of human endometrial regeneration and regression described above was subjected to a BLI-based monitoring system. We first introduced a variant luciferase reporter gene into SDECs using lentivirus, then transplanted the variant luciferase-expressing SDECs beneath the kidney capsule (on the dorsal side) of ovariectomized NOG mice. Eight weeks after xenotransplantation, we observed bioluminescent signals at the transplant sites followed by E_2 treatment (Fig. 9.2d) [60]. We assessed the E_2-dependent growth, P_4 withdrawal-induced breakdown (Fig. 9.2e), and antiestrogen-mediated growth arrest of the endometrium-like tissue generated from lentivirally engineered human endometrial cells in a noninvasive and real-time manner [60]. Thus, by combining transplantation of SDECs with lentivirus-mediated cell engineering, we created an animal model suitable for the study of endometrial physiology/pathophysiology. The model may also be appropriate for the analysis of gene expression in endometrial disorders, drug testing, and various types of neoplastic disease.

9.3.4 Role of Endometrial SP Cells in Endometrial Regeneration as Determined by an In Vivo Endometrial Stem Cell Assay

The characteristics and behaviors of stem cells differ depending on the methods used to isolate, examine, and validate them. In vitro experiments, including clonogenic and differentiation assays, do not necessarily provide an appropriate

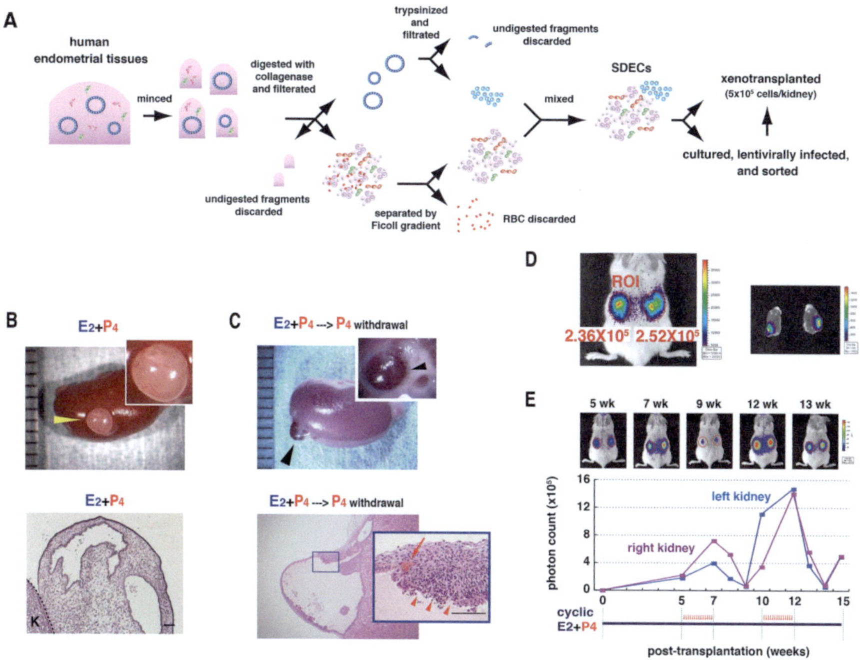

Fig. 9.2 In vivo model of human endometrial regeneration and bioluminescence imaging (Adopted from Masuda et al. [60]). (**a**) Summary of procedures for preparing singly dispersed endometrial cells (SDECs) from human endometrial tissues and representative flow cytometric data on SDECs. (**b**) Macroscopic and hematoxylin and eosin (H&E)-stained tissues from the transplanted site (*arrowhead*) from a NOG mouse treated with estradiol in combination with progesterone (E_2 + P_4) for 10 weeks after xenotransplantation. The borders between the reconstituted tissue and the mouse kidney (K) are indicated by the dotted line. Bar, 100 μm. (**c**) Macroscopic view and H&E-stained tissue from the transplanted site (*arrows*) of a NOG mouse treated with cyclic E_2 + P_4 followed by P_4 withdrawal. The glandular structure was partially disrupted (*arrowheads*), and hemorrhage occurred in the stroma (*arrow*). A *small box* marks a region shown at higher magnification in the adjacent panel as indicated. Bar, 100 μm. (**d**) Optical bioluminescent images and noninvasive quantitative assessment of the endometrial tissues reconstructed from lentivirally transduced SDECs in living NOG mice. A bioluminescence image of the endometrial reconstructs expressing a variant luciferase in a ventrally positioned NOG mouse treated with E_2 alone. Bioluminescent images of the laparotomized mouse at the dorsal position (*lower left*) and its excised kidneys. (**e**) Representative bioluminescent images and serial photon count measurements of xenotransplanted and ovariectomized NOG mice treated with cyclic E_2 + P_4 to induce artificial menstrual cycle-related changes

microenvironment (stem cell niche) required for the maintenance of endometrial stem cell properties. Thus, the characteristics of stem cells often vary between laboratories [14]. Indeed, ESP cells reported in the literature differ with respect to the expression pattern of surface markers, clonal efficiency, preference for culture conditions, and localization in the eutopic normal endometrium [30–34]. These discrepancies are partly attributable to the differences between cESP and fESP as mentioned previously. Furthermore, it is highly likely that the low efficiency of endometrial reconstitution from ESP cells alone is due to the lack of an appropriate microenvironment or stem cell niche to elicit the stem cell potentials of ESP cells, although a cell line derived from freshly isolated SP cells reconstituted endometrial-like tissue at a 100 % success rate [33]. Overall, it remains uncertain whether there are multiple types of stem cells and stem cell niches in the human endometrium and, if so, whether they have distinct roles.

To arrive at a consensus definition of endometrial stem/progenitor cells, we utilized our model of endometrial regeneration described above in combination with the use of cell tracking [54]. ESP and EMP cells were infected with lentivirus carrying tandem tomato (TdTom), a red fluorescent protein (Fig. 9.3a). They were mixed with unlabeled whole endometrial cells and transplanted under the kidney capsule of ovariectomized NOG mice (Fig. 9.3a). These mice were treated with E_2 and P_4 for 8 weeks, after which the kidneys were examined. All of the grafts

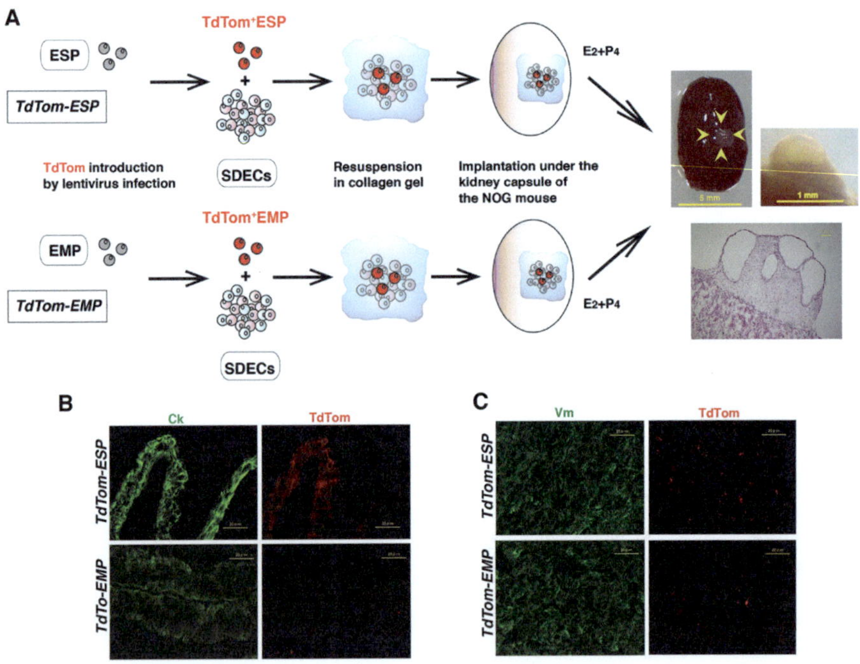

Fig. 9.3 In vivo endometrial stem cell assay (Adopted from Miyazaki et al. [54]). (**a**) Summary of procedures for in vivo endometrial stem cell assay. Macroscopic view and H&E-stained tissue from human endometrium-like tissues reconstituted in the in vivo stem cell assay. Representative macroscopic appearance and H&E-stained tissue from the transplanted site (*arrowheads*) of NOG mice 8 weeks after xenotransplantation of tandem Tomato (TdTom)-endometrial side population (ESP) cells. TdTom, tandem Tomato. (**b**) Representative immunofluorescent images of the TdTom-ESP-derived or TdTom-EMP-derived reconstituted endometrial tissues immunostained with anti-TdTom antibody together with an antibody against vimentin (Vm) or cytokeratin (Ck) as indicated

reconstituted endometrium-like tissues under the kidney capsules (Fig. 9.3a, right three panels). Greater numbers of TdTom-positive cells were found in glandular, stromal, and endothelial tissues of the generated endometrium when ESP cells were transplanted compared to EMP cells (Fig. 9.3b) [54]. These data support the in vivo multi-lineage differentiation potential of ESP. Thus, we have postulated that ESP cells are likely candidate stem cells with a capacity to differentiate into glandular, stromal, endothelial, and smooth muscle cells [54], although it remains possible that multiple stem/progenitor populations are likely present in the ESP fraction. Indeed, investigation of the endometrial SP is hampered by its heterogeneity and low clonogenicity, making single cell culture unrealistic. However, an increasing body of evidence indicates that endometrial stem/progenitor cells are enriched in the SP fraction. In any case, the inclusion of unlabeled, unfractionated endometrial cells in this reconstitution assay provides an appropriate microenvironment or stem cell niche for ESP. This novel in vivo endometrial stem cell assay enables us to track the differentiation of the endometrial stem/progenitor cell candidates into each endometrial lineage. This in vivo endometrial stem cell assay will be useful in determining the nature and roles of endometrial precursors.

9.4 Concluding Remarks

Increasing bodies of evidence indicate that putative stem/progenitor cells exist in the human endometrium and likely play important roles in the regulation of dynamic changes in endometrial tissue homeostasis and function. Many laboratories (including our own) have isolated and/or identified putative endometrial stem/progenitor cells using a variety of methods [14]. Nevertheless, it remains elusive how many types of stem/progenitor cells exist in the human endometrium, their various phenotypes and functions, and whether there is a hierarchical relationship among them. To approach these questions, it is definitely necessary to identify cell surface markers that are specific for stem/progenitor cells in the normal human endometrium. Such data would enable prospective isolation, characterization, and functional analysis of these cells more easily, stably, precisely, and robustly. SSEA-1, CD146, CD140b, and SUS2 have now emerged as promising surface markers [14]. In addition to these markers, supplementing the SP phenotype with specific surface markers remains critically important. The in vivo endometrial stem cell assay that we have developed [54] will be useful not only in further validating these markers but also in determining the nature and roles of endometrial stems/precursors. In accordance with these recent advances in the identification of cell surface markers specific for stem/progenitor cells in the normal human endometrium, research on endometrial stem/progenitor cells now moves from its early phase to explore the molecular and cellular mechanisms by which they are involved in the physiology and pathology of the endometrium.

Acknowledgments I wish to thank Hirotaka Masuda, Kaoru Miyazaki, Masanori Ono, Takashi Kajitani, Hiroshi Uchida, and the other members of my research group for their contributions, assistance, and discussions. I sincerely thank Hideyuki Okano and Yumi Matsuzaki for their generous collaboration. I also thank Rika Shibata for secretarial assistance. This work was partly supported by grant-in-aids from the Japan Society for the Promotion of Science (to T.M and Y.Y.), a grant-in-aid from Keio University Sakaguchi-Memorial Medical Science Fund (to T.M.), and a grant-in-aid from the Japan Medical Association (to T.M.).

References

1. Kahn M. Wnt signaling in stem cells and tumor stem cells. Semin Reprod Med. 2015;33 (5):317–25. doi:10.1055/s-0035-1558404.
2. Weissman IL, Shizuru JA. The origins of the identification and isolation of hematopoietic stem cells, and their capability to induce donor-specific transplantation tolerance and treat autoimmune diseases. Blood. 2008;112(9):3543–53. doi:10.1182/blood-2008-08-078220.
3. Squillaro T, Peluso G, Galderisi U. Clinical trials with mesenchymal stem cells: an update. Cell Transplant. 2015. doi:10.3727/096368915X689622.
4. Dominici M, Le Blanc K, Mueller I, Slaper-Cortenbach I, Marini F, Krause D, et al. Minimal criteria for defining multipotent mesenchymal stromal cells. The International Society for Cellular Therapy position statement. Cytotherapy. 2006;8(4):315–7. doi:10.1080/14653240600855905.
5. Daley GQ. Stem cells and the evolving notion of cellular identity. Philos Trans R Soc Lond Ser B Biol Sci. 2015; 370(1680). doi:10.1098/rstb.2014.0376.
6. Seaberg RM, van der Kooy D. Stem and progenitor cells: the premature desertion of rigorous definitions. Trends Neurosci. 2003;26(3):125–31. doi:10.1016/S0166-2236(03)00031-6.
7. Prianishnikov VA. On the concept of stem cell and a model of functional-morphological structure of the endometrium. Contraception. 1978;18(3):213–23.
8. Padykula HA, Coles LG, Okulicz WC, Rapaport SI, McCracken JA, King Jr NW, et al. The basalis of the primate endometrium: a bifunctional germinal compartment. Biol Reprod. 1989;40(3):681–90.
9. Padykula HA. Regeneration in the primate uterus: the role of stem cells. Ann N Y Acad Sci. 1991;622:47–56.
10. Gargett CE. Uterine stem cells: what is the evidence? Hum Reprod Update. 2007;13 (1):87–101.
11. Tanaka M, Kyo S, Kanaya T, Yatabe N, Nakamura M, Maida Y, et al. Evidence of the monoclonal composition of human endometrial epithelial glands and mosaic pattern of clonal distribution in luminal epithelium. Am J Pathol. 2003;163(1):295–301.
12. Brenner RM, Slayden OD, Rodgers WH, Critchley HO, Carroll R, Nie XJ, et al. Immunocytochemical assessment of mitotic activity with an antibody to phosphorylated histone H3 in the macaque and human endometrium. Hum Reprod. 2003;18(6):1185–93.
13. Gurung S, Deane JA, Masuda H, Maruyama T, Gargett CE. Stem cells in endometrial physiology. Semin Reprod Med. 2015;33(5):326–32. doi:10.1055/s-0035-1558405.
14. Gargett CE, Schwab KE, Deane JA. Endometrial stem/progenitor cells: the first 10 years. Hum Reprod Update. 2015. doi:10.1093/humupd/dmv051.
15. Masuda H, Maruyama T, Gargett CE, Miyazaki K, Matsuzaki Y, Okano H, et al. Endometrial side population cells: potential adult stem/progenitor cells in endometrium. Biol Reprod. 2015;93(4):84. doi:10.1095/biolreprod.115.131490.
16. Maruyama T. Endometrial stem/progenitor cells. J Obstet Gynaecol Res. 2014;40(9):2015–22. doi:10.1111/jog.12501.

17. van Os R, Kamminga LM, de Haan G. Stem cell assays: something old, something new, something borrowed. Stem Cells. 2004;22(7):1181–90.
18. Chan RW, Schwab KE, Gargett CE. Clonogenicity of human endometrial epithelial and stromal cells. Biol Reprod. 2004;70(6):1738–50.
19. Gargett CE, Schwab KE, Zillwood RM, Nguyen HP, Wu D. Isolation and culture of epithelial progenitors and mesenchymal stem cells from human endometrium. Biol Reprod. 2009;80 (6):1136–45. doi:10.1095/biolreprod.108.075226.
20. Valentijn AJ, Palial K, Al-Lamee H, Tempest N, Drury J, Von Zglinicki T, et al. SSEA-1 isolates human endometrial basal glandular epithelial cells: phenotypic and functional characterization and implications in the pathogenesis of endometriosis. Hum Reprod. 2013;28 (10):2695–708. doi:10.1093/humrep/det285.
21. Schwab KE, Gargett CE. Co-expression of two perivascular cell markers isolates mesenchymal stem-like cells from human endometrium. Hum Reprod. 2007;22(11):2903–11. doi:10.1093/humrep/dem265.
22. Schwab KE, Hutchinson P, Gargett CE. Identification of surface markers for prospective isolation of human endometrial stromal colony-forming cells. Hum Reprod. 2008;23 (4):934–43. doi:den051 [pii] 10.1093/humrep/den051.
23. Masuda H, Anwar SS, Buhring HJ, Rao JR, Gargett CE. A novel marker of human endometrial mesenchymal stem-like cells. Cell Transplant. 2012;21(10):2201–14. doi:10.3727/096368911X637362.
24. Spitzer TL, Rojas A, Zelenko Z, Aghajanova L, Erikson DW, Barragan F, et al. Perivascular human endometrial mesenchymal stem cells express pathways relevant to self-renewal, lineage specification, and functional phenotype. Biol Reprod. 2012;86(2):58. doi:10.1095/biolreprod.111.095885.
25. Murakami K, Lee YH, Lucas ES, Chan YW, Durairaj RP, Takeda S, et al. Decidualization induces a secretome switch in perivascular niche cells of the human endometrium. Endocrinology. 2014;155(11):4542–53. doi:10.1210/en.2014-1370.
26. Lucas ES, Dyer NP, Murakami K, Lee YH, Chan YW, Grimaldi G, et al. Loss of endometrial plasticity in recurrent pregnancy loss. Stem Cells. 2015. doi:10.1002/stem.2222.
27. Zhou S, Schuetz JD, Bunting KD, Colapietro AM, Sampath J, Morris JJ, et al. The ABC transporter Bcrp1/ABCG2 is expressed in a wide variety of stem cells and is a molecular determinant of the side-population phenotype. Nat Med. 2001;7(9):1028–34.
28. Challen GA, Little MH. A side order of stem cells: the SP phenotype. Stem Cells. 2006;24 (1):3–12.
29. Goodell MA, Brose K, Paradis G, Conner AS, Mulligan RC. Isolation and functional properties of murine hematopoietic stem cells that are replicating in vivo. J Exp Med. 1996;183 (4):1797–806.
30. Masuda H, Matsuzaki Y, Hiratsu E, Ono M, Nagashima T, Kajitani T, et al. Stem cell-like properties of the endometrial side population: implication in endometrial regeneration. PLoS ONE. 2010;5(4), e10387. doi:10.1371/journal.pone.0010387.
31. Kato K, Yoshimoto M, Kato K, Adachi S, Yamayoshi A, Arima T, et al. Characterization of side-population cells in human normal endometrium. Hum Reprod. 2007;22(5):1214–23.
32. Tsuji S, Yoshimoto M, Takahashi K, Noda Y, Nakahata T, Heike T. Side population cells contribute to the genesis of human endometrium. Fertil Steril. 2008;90(4 Suppl):1528–37. doi:10.1016/j.fertnstert.2007.08.005.
33. Cervello I, Mas A, Gil-Sanchis C, Peris L, Faus A, Saunders PT, et al. Reconstruction of endometrium from human endometrial side population cell lines. PLoS ONE. 2011;6(6), e21221. doi:10.1371/journal.pone.0021221 PONE-D-11-04638 [pii].
34. Cervello I, Gil-Sanchis C, Mas A, Delgado-Rosas F, Martinez-Conejero JA, Galan A, et al. Human endometrial side population cells exhibit genotypic, phenotypic and functional features of somatic stem cells. PLoS ONE. 2010;5(6), e10964. doi:10.1371/journal.pone.0010964.

35. Arai F, Hirao A, Ohmura M, Sato H, Matsuoka S, Takubo K, et al. Tie2/angiopoietin-1 signaling regulates hematopoietic stem cell quiescence in the bone marrow niche. Cell. 2004;118(2):149–61.
36. Taylor HS. Endometrial cells derived from donor stem cells in bone marrow transplant recipients. JAMA. 2004;292(1):81–5.
37. Bratincsak A, Brownstein MJ, Cassiani-Ingoni R, Pastorino S, Szalayova I, Toth ZE, et al. CD45-positive blood cells give rise to uterine epithelial cells in mice. Stem Cells. 2007;25(11):2820–6. doi:10.1634/stemcells.2007-0301.
38. Ikoma T, Kyo S, Maida Y, Ozaki S, Takakura M, Nakao S, et al. Bone marrow-derived cells from male donors can compose endometrial glands in female transplant recipients. Am J Obstet Gynecol. 2009;201(6):608 e1–8. doi:10.1016/j.ajog.2009.07.026.
39. Cervello I, Gil-Sanchis C, Mas A, Faus A, Sanz J, Moscardo F, et al. Bone marrow-derived cells from male donors do not contribute to the endometrial side population of the recipient. PLoS ONE. 2012;7(1), e30260. doi:10.1371/journal.pone.0030260.
40. Kim JY, Tavare S, Shibata D. Counting human somatic cell replications: methylation mirrors endometrial stem cell divisions. Proc Natl Acad Sci U S A. 2005;102(49):17739–44.
41. Hida N, Nishiyama N, Miyoshi S, Kira S, Segawa K, Uyama T, et al. Novel cardiac precursor-like cells from human menstrual blood-derived mesenchymal cells. Stem Cells. 2008;26 (7):1695–704. doi:10.1634/stemcells.2007-0826.
42. Zhang Y, Huang B. Peripheral blood stem cells: phenotypic diversity and potential clinical applications. Stem Cell Rev. 2012;8(3):917–25. doi:10.1007/s12015-012-9361-z.
43. Henriet P, Gaide Chevronnay HP, Marbaix E. The endocrine and paracrine control of menstruation. Mol Cell Endocrinol. 2012;358(2):197–207. doi:10.1016/j.mce.2011.07.042.
44. Novak E, Te linde RW. The endometrium of the menstruating uterus. JAMA. 1924;83:900–6.
45. Ferenczy A. Studies on the cytodynamics of human endometrial regeneration. I. Scanning electron microscopy. Am J Obstet Gynecol. 1976;124(1):64–74.
46. Ludwig H, Metzger H, Frauli M. Endometrium: tissue remodelling and regeneration. In: d'Arcangues C, Fraser IS, Newton JR, Odlind V, editors. Contraception and mechanisms of endometrial bleeding. Cambridge: Cambridge University Press; 1990. p. 441–66.
47. Salamonsen LA. Tissue injury and repair in the female human reproductive tract. Reproduction. 2003;125(3):301–11.
48. Maybin J, Critchley H. Repair and regeneration of the human endometrium. Expert Rev Obstet Gynecol. 2009;4:283–98.
49. Garry R, Hart R, Karthigasu KA, Burke C. A re-appraisal of the morphological changes within the endometrium during menstruation: a hysteroscopic, histological and scanning electron microscopic study. Hum Reprod. 2009;24(6):1393–401. doi:10.1093/humrep/dep036.
50. Garry R, Hart R, Karthigasu KA, Burke C. Structural changes in endometrial basal glands during menstruation. BJOG. 2010;117(10):1175–85. doi:10.1111/j.1471-0528.2010.02630.x.
51. Gaide Chevronnay HP, Galant C, Lemoine P, Courtoy PJ, Marbaix E, Henriet P. Spatiotemporal coupling of focal extracellular matrix degradation and reconstruction in the menstrual human endometrium. Endocrinology. 2009;150(11):5094–105. doi:10.1210/en.2009-0750.
52. Huang CC, Orvis GD, Wang Y, Behringer RR. Stromal-to-epithelial transition during postpartum endometrial regeneration. PLoS ONE. 2012;7(8), e44285. doi:10.1371/journal.pone.0044285.
53. Patterson AL, Zhang L, Arango NA, Teixeira J, Pru JK. Mesenchymal-to-epithelial transition contributes to endometrial regeneration following natural and artificial decidualization. Stem Cells Dev. 2013;22(6):964–74. doi:10.1089/scd.2012.0435.
54. Miyazaki K, Maruyama T, Masuda H, Yamasaki A, Uchida S, Oda H, et al. Stem cell-like differentiation potentials of endometrial side population cells as revealed by a newly developed in vivo endometrial stem cell assay. PLoS ONE. 2012;7(12), e50749. doi:10.1371/journal.pone.0050749.

55. Maruyama T, Masuda H, Ono M, Kajitani T, Yoshimura Y. Human uterine stem/progenitor cells: their possible role in uterine physiology and pathology. Reproduction. 2010;140 (1):11–22. doi:REP-09-0438 [pii] 10.1530/REP-09-0438.
56. Sasson IE, Taylor HS. Stem cells and the pathogenesis of endometriosis. Ann N Y Acad Sci. 2008;1127:106–15.
57. Maruyama T, Yoshimura Y. Stem cell theory for the pathogenesis of endometriosis. Front Biosci. 2012;4:2854–63.
58. Maruyama T. Role of stem cells in the pathogenesis of endometriosis. In: Harada T, editor. Endometriosis: pathogenesis and treatment. Tokyo: Springer Japan; 2014. p. 33–48.
59. Hufnagel D, Li F, Cosar E, Krikun G, Taylor HS. The role of stem cells in the etiology and pathophysiology of endometriosis. Semin Reprod Med. 2015;33(5):333–40. doi:10.1055/s-0035-1564609.
60. Masuda H, Maruyama T, Hiratsu E, Yamane J, Iwanami A, Nagashima T, et al. Noninvasive and real-time assessment of reconstructed functional human endometrium in NOD/SCID/γ_c^{null} immunodeficient mice. Proc Natl Acad Sci U S A. 2007;104(6):1925–30.
61. Masuda H, Okano HJ, Maruyama T, Yoshimura Y, Okano H, Matsuzaki Y. In vivo imaging in humanized mice. Curr Top Microbiol Immunol. 2008;324:179–96.

MIX
Papier aus verantwortungsvollen Quellen
Paper from responsible sources
FSC® C105338

If you have any concerns about our products,
you can contact us on
ProductSafety@springernature.com

In case Publisher is established outside the EU,
the EU authorized representative is:
Springer Nature Customer Service Center GmbH
Europaplatz 3, 69115 Heidelberg, Germany

Printed by Libri Plureos GmbH
in Hamburg, Germany